Wastewater Treatment Using Advanced Oxidation Processes

Wastewater Treatment Using Advanced Oxidation Processes

Edited by

Dr. Saeed Farrokhpay

ARCLER
PRESS

www.arclerpress.com

Wastewater Treatment Using Advanced Oxidation Processes

Dr. Saeed Farrokhpay

Arcler Press
4164 Lakeshore Road
Burlington, ON L7L 1A4
Canada
www.arclerpress.com
Email: orders@arclereducation.com

ISBN: 978-1-77956-436-8 (Hardcover)

ABOUT THE EDITOR

Dr Saeed Farrokhpay is a Chemical Engineer with several years of experience in mineral & material processing. He obtained his PhD from University of South Australia in 2005. He is currently a Technical Consultant in Australia. He has worked for more than 20 years at mineral and chemical industries, universities and research centers around the world. Dr Farrokhpay has published more than 90 papers in high ranked journals and conference proceedings. He has also edited several technical and scientific books, and served as an editorial board member of several international scientific journals.

TABLE OF CONTENTS

LIST OF FIGURES

LIST OF TABLES

PREFACE

Water pollution is a major contributor to the world's growing pollution problem. Hazardous organic waste is mostly generated by industries, households, hospitals, and municipalities. These wastes represent one of the greatest issues for environmental engineers. Rapid industrialization, transportation, and urbanization, combined with population increase and deforestation, are putting a strain on freshwater supplies worldwide. Currently utilized wastewater treatment systems have downsides. There is a dire need to develop an eco-friendly technology for wastewater treatment. Advanced Oxidation Processes (AOPs) have a role in this scenario. AOPs are water treatment techniques that generate enough hydroxyl radicals at near-ambient temperatures and pressures. The hydroxyl radical is a powerful, non-selective, green chemical oxidant that quickly destroys organic pollutants, including recalcitrant compounds. Various wastewater treatment methods, including photocatalytic oxidation, sonolysis, photo-Fenton processes, catalytic wet peroxide oxidation, microwave/hydrogen peroxide, supercritical water oxidation, ozone-based processes, X-ray, γ-ray, and electron beam-based processes, have demonstrated efficiency at pilot scale. This technology will soon be useful in tackling global water pollution.

This book comprises of eight-chapters; the general composition, distribution, treatment, and management of water are identified in the first chapter. Additionally, the goals and principles of water and wastewater treatment are also examined in this chapter. In the second chapter, the fundamental mechanisms of different advanced oxidation processes, such as ozonation, hydrogen peroxide, UV, persulfate, and Fenton oxidation, are summarized.

The third chapter is about the processes based on ozone for wastewater treatment which highlights the ozonation of wastewater and the by-products obtained in this process. The heterogeneous Fenton-like processes are discussed in the fourth chapter of this book, discussing the different types of chemical and continuous reactors along with Fenton's oxidation.

The fifth chapter describes the process of photo catalysis and discusses the different types of semiconductor photo catalysts used in photo catalysis. The various applications of photo catalysis used in wastewater treatment are also part of this chapter. The sixth chapter deals with the fundamentals of photo-activated persulfate-based advanced oxidation processes for wastewater treatment.

Electrochemical oxidation processes along with their advantages, drawbacks, applications, and prospects are discussed in chapter seven. The final chapter navigates through the applications of ultraviolet-based advanced oxidation processes in wastewater treatment.

This book targets scientists, researchers, students, engineers, professionals, and practitioners from prominent organizations and corporations, including water associations, wastewater management companies, solid waste disposal operators, and lawmakers.

CHAPTER 1

INTRODUCTION TO WASTEWATER TREATMENTS

LEARNING OBJECTIVES

At the end of this chapter, readers will be able to:

1. Comprehend the critical importance of water quality concerning public health and environmental sustainability.
2. Understand the significance of water and identify sources of wastewater.
3. Gain insight into the historical development of wastewater treatment.
4. Learn about the different types of wastewater treatments.
5. Comprehend the limitations associated with different wastewater treatments.

INTRODUCTORY EXAMPLE

Care Pharmaceuticals, a leading player in the pharmaceutical industry, faces a significant challenge in managing the wastewater generated during its manufacturing processes. With stringent environmental regulations and a commitment to sustainability, they recognize the imperative of treating the wastewater effectively to mitigate impact of environment and ensure compliance with regulatory standards.

The wastewater produced by Care Pharmaceuticals comprises a complex mixture of inorganic and organic contaminants. To address this challenge, Care Pharmaceuticals invests in state-of-the-art treatment methods.

Through the implementation of advanced treatment solutions, Care Pharmaceuticals achieves efficient removal of pollutants, thereby reducing the environmental footprint of its operations and safeguarding the health and well-being of surrounding communities. This strategic approach not only aligns with the company's corporate social responsibility initiatives but also reinforces its reputation as an environmentally responsible industry leader.

UNIT INTRODUCTION

On the planet, there is an astonishing volume of water – approximately 332.5 million cubic miles. However, it is important to note that less than 1% of this vast resource is readily available for everyday use, such as for households and industries. Freshwater resources are commonly seen as an unlimited supply due to the natural process of water purification and replenishment by means of the hydrologic cycle (Parker, 2011). However, this is not necessarily the case if water reserves continue to be polluted or depleted at a faster rate than they can be replenished. In addition, approximately two-thirds of the available water is safe for drinking, while the remaining portion is contaminated and unfit for consumption due to pollution and various other factors. The depletion of water resources is a pressing global issue, highlighting the need for accessible and safe water to ensure the well-being of communities. By raising public awareness and taking proactive measures, humans can prevent environmental issues and safeguard against waterborne illnesses. The treatment of wastewater is essential for addressing contamination. The treatment technology mirrors the natural water cycle with the goal of cleansing, purifying, and restoring water to meet drinking water standards and regulatory requirements (Gedda et al., 2021).

Did You Know?

The inaugural municipal wastewater treatment facility was constructed in the city of London in 1868. The system employed a combination of biological and physical mechanisms to process sewage and enhance the water quality of the River Thames.

1.1. WATER AND DISEASE

Learning Objectives

- Gain knowledge about threatening diseases caused by contaminated water.

Illness is a specific abnormal condition that impacts a portion or the entirety of an organism. Illness can arise from more than just physical harm. It can be triggered by external elements like pathogens or internal issues like immune system dysfunctions, such as autoimmune disorders or hypersensitivity. Polluted water can lead to various diseases worldwide, some of which can be life-threatening. Waterborne disease is a global health issue, as highlighted by the World Health Organization (WHO) (Ashbolt, 2004). The focus is on individuals with weakened immune systems, such as young children, the elderly, and those who are ill. In developing countries, a significant number of cases has been reported with a staggering 80% of illnesses being attributed to water-borne diseases. Among children, dehydration caused by diarrhea is the leading cause of mortality. Water, despite being considered a renewable resource, is a precious and limited commodity that should be valued and protected. The presence of hazardous bacteria is frequently associated with contaminated water, which can result in the transmission of diseases. These diseases are commonly referred to as water-borne illnesses. However, it is significant to consider other significant factors, including different chemicals such as organic compounds, inorganic compounds, and radionuclides. These chemicals can enter the water through inadequate waste management, subpar agricultural practices, and unregulated mining. Several of these substances pose minimal risk to human health due to their low concentrations. Nevertheless, certain substances have been found to have significant health impacts even at low concentrations, necessitating the implementation of treatment methods to eliminate or minimize their presence in drinking water. There is a potential for infection through both direct and indirect exposures related to these two categories of contaminants. Consuming water contaminated with pathogenic microorganisms can lead to direct infection caused by biological factors. Examples of these microorganisms include Salmonella typhi, Campylobacter spp., Shigella spp., Vibrio cholera, and others. On the other hand, the indirect approach relies on the presence of vectors in the contaminated water (Nichols et al., 2009). These vectors can be found in aquatic environments or rely on water for their reproductive processes. As an illustration, some vectors that can transmit water-borne diseases include ticks, snails, mosquitoes, flies, and fleas. Chemical-induced diseases can occur when individuals consume contaminated water containing harmful substances like heavy metals, nitrates, pesticides, and disinfection by-products. One way to indirectly ingest harmful chemicals is by consuming marine life, as they tend to accumulate these substances in their bodies. When people consume contaminated organisms, toxic substances can build up in their bodies, leading to the development of specific symptoms associated with the contaminants (Sontrop et al., 2013).

1.2. WATER FOR LIFE

Learning objectives

- Understand the significance of water for sustainability of life on earth.
- Comprehend the composition of water.

Water is essential for sustaining and nourishing life on Earth. Water plays a crucial role in the complex mechanisms of the planet's climate and its interaction with the atmosphere, land, and oceans. In addition, water can resupply and distribute itself through a process known as the hydrologic cycle. Water is a versatile resource that exists in liquid, gaseous and solid states on the Earth's surface. From a chemical perspective, water is an inorganic substance made up of H_2O molecules and is a crucial component of all living organisms (Mahgoub et al., 2010). Water possesses distinct properties that contribute to its importance for living organisms. It acts as a biochemical solvent, regulates temperature, and provides a habitat for various life forms. In addition, water has a polar molecular structure with a positive terminal on hydrogen and a negative on oxygen. Water has the unique ability to dissolve a variety of substances. Also, water possesses unique properties that enable it to maintain a favorable temperature range for various living organisms.

As an illustration, the amount of energy needed for a gram of water to raise its internal temperature by 1°C is 4.2 kJ/g/°C. Given the composition of cells, a significant amount of energy is needed to significantly increase their temperature. This ensures that the surrounding temperature remains stable, enabling optimal conditions for biochemical reactions to take place (Sturm et al., 2017). The process of sweating enables the body to dissipate heat by absorbing it and subsequently eliminating it. In cold climates, the absence of ice formation until a significant amount of heat escapes from water enables the existence of aquatic organisms due to the very high latent heat of fusion. These water properties support a range of natural processes that occur in the environment. Water serves as a habitat for various aquatic organisms. The existence of dissolved oxygen in water is crucial for the respiration of aquatic life, just like their counterparts on land. Fish rely on their specialized gills to extract oxygen from the water, while aquatic vegetation and phytoplankton need dissolved oxygen for photosynthesis when sunlight is not accessible. Underneath the floating ice, fish and crustaceans in cold climates can survive due to the lower density of ice compared to liquid water. The presence of floating ice serves as a protective barrier, preventing the lower portion of the lake from freezing entirely in the face of harsh, frigid temperatures. The cohesive force of water allows small insects like striders to effortlessly walk on its surface (Galiani et al., 2005).

1.3. SOURCES OF WASTEWATER

Learning Objectives:

- Recognize the different sources of wastewater including household, industry, and animal wastes.

Wastewater is a type of water that becomes contaminated as a result of various human activities. The composition of wastewater can vary significantly across different regions due to varying factors. Municipal wastewater is the combination of all sources of water, excluding stormwater runoff. The liquid and solid discharge from animal and human excretion systems, particularly the gastrointestinal system, is comprised of human and animal wastes. The organic waste is composed of water, fats, protein, biomass, and fibers (Petrie, 2021). Contaminated water contains a multitude of harmful microorganisms that have the potential to cause serious illnesses. In addition to emitting a potent and unpleasant smell, the water polluted by this source poses a significant risk to the environment and food sources. Household waste contributes to the generation of wastewater.

The composition of wastewater can include substances such as soap, grease, food particles, and waste products from various sources like sinks, toilets, showers, and washing machines. It is important to consider the impact of certain elements found in household waste on the delicate balance of the natural ecosystem. Additionally, the presence of dangerous microbes can pose a risk to public health. Typically, domestic wastewater contains a small percentage of impurities, less than 0.2%. The wastewater in question falls under the category of low-strength wastewater, typically with a COD level below 2,000 mg/L (Lee et al., 2016).

Industrial waste often contains high concentrations of metallic substances and toxic chemicals that are specific to the industry.

This waste can have extremely high levels of pH, temperature, and excess nutrients, which can harm the environment. Industrial wastewater treatment typically involves a dechlorination process, addressing recalcitrant water, and eliminating hazardous substances from the waste effluent (Shete & Shinkar, 2013) (Figure 1.1).

Remember

Wastewater can come from different sources, including domestic sewage, industrial effluents, and agricultural runoff. Each source presents unique challenges for treatment and management.

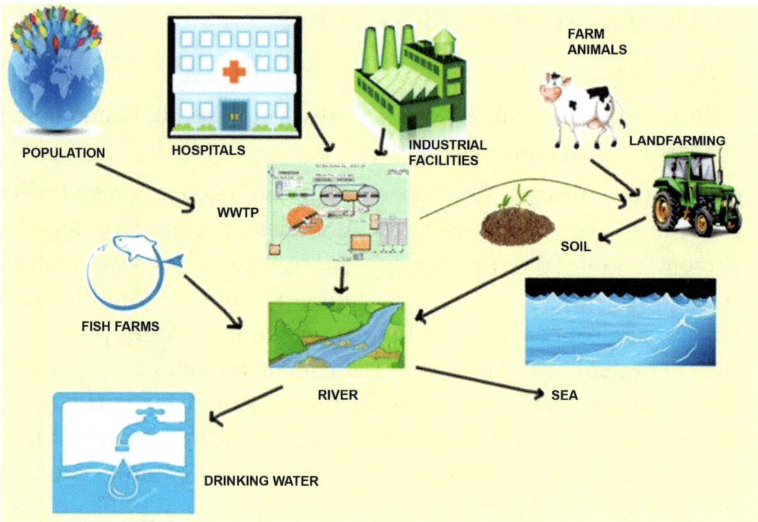

Figure 1.1. *Representation of the sources of wastewater.*

Source: Mansi Rastogi, Creative Commons License.

1.4. HISTORY OF WASTEWATER TREATMENT

Learning Objectives:

- Understand the significance of treatment of wastewater.
- Comprehend the historical development of wastewater treatment.

The treatment of water has been valued for centuries due to its significant role in maintaining water quality. Since ancient times, the inhabitants of India have employed charcoal for water filtration and copper pots for water preservation. It is believed that the ancient Egyptians, dating back to 1500 BC, utilized innovative methods such as filtration using coagulation and wick siphon with chemical alum to enhance the clarity of water. Alternative methods, including boiling, utilizing hot metals, and sand filtration, have long been suggested as effective treatment options for centuries. In 500 BC, Hippocrates discovered the curative qualities of water and created a water filter called the Hippocrates sleeve.

The filter was designed to minimize sediments that may lead to undesirable color and odor. During the post-classical period, there was a limited understanding of water treatment methods. Francis Bacon proposed an experiment where he dug a trench near a seashore, to filter seawater through layers of sand to obtain salt-free, pure water. Despite the failure of his hypothesis, he expressed enthusiasm for future water filtration research (Xu et al., 2020). During the 1700s, an early form of domestic water filtration was developed using materials such as wool, sponge, and charcoal. In 1804, Robert Thom made a significant contribution to the field of water purification by designing a prototype water purification plant in Scotland. Water underwent treatment through the process of slow filtering with sand and was subsequently distributed using a horse-drawn van. In 1849, London experienced a devastating epidemic, which left a lasting impact on the city. According to a study lead by Streiner et al. (2010), the cholera outbreak resulted in a significant loss of life, with 14,137 fatalities and numerous cases of infection.

In contrast, it was only in 1854 that John Snow successfully demonstrated that cholera was transmitted by polluted water, rather than through miasma. His notable work also included the incorporation of chlorine to purify water contaminated with cholera. After achieving impressive results, the British government introduced chlorination sand filter methods for public use. Consequently, it led to the creation of the world's first government water regulation (Spielvogel, 2007). Since 1914, the United States implemented drinking water standards to regulate the bacterial count in drinking water supplies. Subsequently, the standard set by the Public Health Service underwent multiple revisions, with minor adjustments made over time. However, in the

1960s, it became evident that the rapid expansion of industrial and agricultural sectors had led to the emergence of numerous xenobiotic compounds. These newly developed synthetic substances can infiltrate water sources through the discharge from industrial facilities, runoff from agricultural fields, and leakage from disposal tanks. Despite the availability of various treatment methods like aeration, flocculation, and activated carbon adsorption, their utilization and effectiveness in removing these emerging chemical contaminants were limited. There has been a growing awareness of the presence of chemical contamination, which has resulted in the implementation of the Safe Drinking Water Act in 1974. During 1980s, there were significant advancements in membrane formation for reverse osmosis filtration, alongside the growing popularity of disinfectants like ozone and chlorine. Certain treatments were motivated by the existence of microorganisms that are resistant to chlorine and can lead to various diseases. Currently, many treatment plants continue to use traditional methods that have been in practice for thousands of years. These methods include coagulation, filtration, and chlorination. However, some modern water treatment plants have adopted more advanced techniques like advanced oxidation, membrane filtration, and ion exchange (Kroiss et al., 2021).

1.5. TYPES OF WATER AND WASTEWATER TREATMENT

Learning Objectives:

- Understand the physical methods for treatment of wastewater.
- Comprehend the advanced oxidation process for wastewater treatment.
- Learn about biological and chemical treatments of wastewater.

Proper treatment is essential for discharges of wastewater and drinking water to ensure its safety for consumption or release. The processes involved in water treatment play a crucial role in maintaining the quality of water to meet the standards set by the WHO for drinking water. For wastewater treatment, guidelines are generally issued by the Ministry of Environment or the municipality in each country. These guidelines are usually categorized into two standards: A and B. These two standards deal with the discharge of wastewater. Standard A pertains to the release of wastewater before it reaches the watershed or the treatment plant whereas Standard B deals with the discharge of wastewater after it has gone through the plant, taking into account the stream flow. Water treatment can be categorized into different processes, including, chemical, physical, and biological treatments. Having a solid grasp of the process, designs, diagrams, and specifications is essential during the initial phase of selecting a wastewater and water treatment method. Understanding the interconnectedness of different processes and their proper sequencing is crucial for achieving efficient treatment (Kataki et al., 2021).

1.5.1. Physical Treatments

The screen, the clarifier, and the grit chamber are all examples of physical treatment units. These units remove impurities through the use of physical processes such as mechanical straining and gravity settling mechanisms. Screens do the mechanical straining process, whereas clarifiers do the gravity-settling process. It is also possible to refer to clarifiers as sedimentation basins or settling tanks. The removal of coarse particles and other bulky substances that are frequently present in raw wastewater is the task that is intended to be accomplished by preliminary treatment (Noor et al., 2023).

1.5.1.1. Screening

It is the process of removing larger particles that have the potential to prevent subsequent equipment or processes from functioning properly. Based on the findings of Gogate & Pandit (2004), screening systems can be classified broadly as either coarse or fine. High number of screenings and contaminants are produced by openings of a smaller size. It is necessary to handle, store, and dispose of the material that has been screened. To design the screening, it is necessary to take careful measurements of the grid spacing, the materials used in construction, and the dimensions. In addition to that, the depth and width of channel, and the approach velocity are also among the factors that are taken into consideration (Burger et al., 2011).

1.5.1.2. Grit Removal

Grit removal is a service that is typically offered at all treatment plants. This service is used to remove materials that are not putrescible and are denser than organic matter. The moving machinery and pumps are shielded from excessive wear and abrasion by this, which in turn prevents the pipes from becoming clogged. Additionally, it serves to prevent the cementing effect from occurring on the bottom of the tanks, while simultaneously reducing the buildup of inert materials within the tanks (Finger & Parrick, 1980).

1.5.1.3. Velocity-Controlled Grit Channel (Horizontal Flow)

This method is considered to be a classic approach for eliminating grit from the surface. In order to ensure sufficient time for the sediment particles to settle, it is advisable to maintain a velocity that is approximately between 0.25 and 0.35

meters per second. The parabolic shape will produce the most favorable flow pattern, and the channel's length will be determined by the depth that is needed by the settling velocity (Rezai & Allahkarami, 2021) (Figure 1.2).

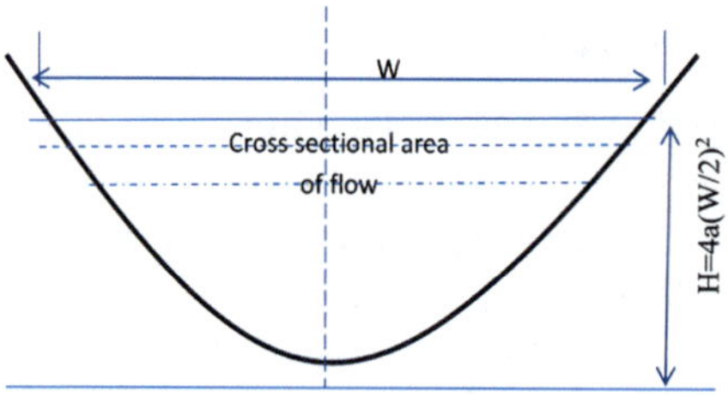

Figure 1.2. *Illustration of ideal parabolic cross-section of grit chamber.*

Source: Husnul Azan, Creative Commons License.

1.5.2. Advanced Oxidation Processes (AOPs)

The advanced oxidation processes (AOPs) have been widely studied for the treatment of drinking water and industrial effluents. Generally, most AOPs are based on the generation of the highly reactive oxidizing species, such as hydroxyl radicals, which could degrade a variety of organic pollutants. Advanced oxidation processes (AOPs) are currently popular in effluent treatment to degrade recalcitrant compounds, which are difficult to remove by biological processes. Despite being highly efficient, when applied in an isolated way in leachate treatment, AOP do not reach effluent discharge standards established by the existing legislation, requiring post-treatment techniques to also be applied. They are also characterized by a little selectivity of attack which is a useful attribute for an oxidant used in wastewater treatment and for solving pollution problems. The versatility of AOP is also enhanced by the fact that they offer different possible ways for radicals' production. Thus allowing a better compliance with the specific treatment requirements. A suitable application of AOP to solid waste leachate treatments must consider that they make use of expensive reactants as H_2O_2, and/or O_3. Therefore, it is obvious that their application should not replace, whenever possible, the more economic treatments as the biological degradation.

The potentialities offered by AOP can be exploited to integrate biological treatments by an oxidative degradation of toxic or refractory substances entering or leaving the biological stage. Another aspect concerning the opportunity of AOP application is that referring to the polluting load of wastes normally expressed as COD. Only wastes with relatively small COD contents (≤5.0 g/l) can be suitably treated by means of these techniques since higher COD contents would require the consumption of too large amounts of expensive reactants. Wastes with more massive pollutants contents can be more conveniently treated by means of wet oxidation or incineration.

The biological treatment process can be significantly complicated by several substances that are present in wastewater from both commercial and municipal sources, regardless of whether these substances are produced artificially or occur naturally. These chemicals possess both great resistance to degradation and the potential to harm the microorganisms involved. The advanced oxidation process is a treatment method which makes use of the power of oxidizing agents or irradiation to efficiently eradicate both inorganic and organic pollutants from the environment. The first step in this process involves transforming complex molecules into compounds that are easier to disassemble. This transition takes place during the first stage. The increase in the production of biodegradable substances will lead to a microbial process that is more efficient and productive as a result. Because of this, the advanced oxidation process is frequently utilized as an additional method to enhance the effectiveness of conventional biological treatment (Deng & Zhao, 2015).

Practice Problem

A wastewater treatment plant is considering the implementation of Advanced Oxidation Processes (AOPs) to remove organic contaminants from effluent. If the AOPs have a removal efficiency of 95% for organic pollutants, calculate the concentration of residual pollutants in the effluent after treatment if the influent concentration is 100 mg/L.

Solution to Practice Problem

Concentration of residual pollutants in effluent = Influent concentration * (1 – Removal efficiency)

= 100 mgL^{-1} * (1–0.95)

= 100 mgL^{-1} * 0.05

= 5 mg^{L-1}

1.5.3. Biological Treatments

Biological treatment, which is also referred to as secondary treatment, utilizes biological processes in the secondary sedimentation phase. Wastewater treatment involves the removal of suspended solids using microorganisms like fungi, algae, or bacteria. This process occurs in the existence of either oxygen (aerobic) or the absence of oxygen (anaerobic), during which the organic matter existing in the wastewater undergoes oxidation or is converted into cells. These cells can subsequently be eradicated through sedimentation or alternative elimination procedures. The treated wastewater is directed into a specialized bioreactor where microorganisms are employed to break down the organic matter under either aerobic or anaerobic conditions. The bioreactor is purposefully engineered to establish an ideal habitat for microorganisms to flourish and harness organic matter as an energy source.

The microorganisms receive oxygen and nourishment from settled wastewater, which allows them to continue oxidizing dissolved organic matter. The biological process mainly revolves around bacteria that play a crucial role in the food chain within the bioreactor. The process of converting organic matter into bacterial biomass is extremely efficient in the purification of wastewater.

Subsequently, it is crucial to eliminate the microbial biomass from the treated water through sedimentation. The secondary sedimentation process closely resembles the primary sedimentation process, differing only in the composition of the sludge, which primarily comprises bacterial cells instead of fecal solids. Heterotrophic bacteria play a critical role in the biological removal of organic material from settled wastewater. These microbes use biological oxidation and biosynthesis to decompose organic materials. Some byproducts are produced during the biological oxidation process and gathered in the water solution before being discharged with the treated effluent (Eqn. (1)). A denser biomass that can be eliminated by sedimentation is created during the biosynthesis phase when the suspended and dissolved materials is transformed into new cells (Eqn. (2)) (Oller et al., 2011).

Oxidation:

$$\text{COHNS} + O_2 + \text{Bacteria} \rightarrow CO_2 + NH_3 + \text{other end product} \quad (1)$$

(Organic matter)

Biosynthesis:

$$\text{COHNS} + O_2 + \text{Bacteria} \rightarrow C_5H_7NO_2 \quad (2)$$

(Organic matter)

1.5.3.1. Biological Treatment Principles

In the biological treatment process, it is important to consider certain core principles:

- Severe changes in hydraulic loads can greatly impact the sensitivity of biological systems. Variations exceeding 250% in a single day can pose challenges as they may lead to a significant reduction in biomass within the clarifiers.
- The growth pace of microorganisms is influenced by temperature. A decrease of 10°C in the temperature of wastewater will significantly reduce the rates of biological reactions by half (Samer, 2015).
- Achieving up to 95% efficiency in BOD removal is possible, although it may not be feasible to eradicate the entire BOD due to the requirement of large tanks.
- It appears to be quite challenging for biological treatment systems to effectively manage sudden increases in waste loads. Equalization is a crucial step when there is a significant fluctuation in the strength of the wastewater or when the wastewater reaches a peak concentration of over 1,000 $mg.L^{-1}$ BOD.
- The C:N:P ratio of wastewater is typically optimal. For the most advantageous biological process, it is recommended that this ratio of commercial wastewater falls within the range of 100:20:1 to 100:5:1.
- Before biological treatment, it is necessary to pretreat oils and solids to prevent any negative impact on the efficiency of wastewater treatment.
- Materials that are poisonous or biologically resistant need special attention since they might require pretreatment before being added to a biological treatment system (Augustos & Von Sperling, 2005).

1.5.3.2. Aerobic Treatment

Investigating the realm of aerobic treatment Historically, aeration has been utilized to eradicate small amounts of volatile organic compounds (VOCs) in water. It has also been used in the process known as "gas adsorption" or "oxidation" to move a material, like oxygen, from the air or a gaseous condition into the water. This process is specifically utilized to oxidize iron and manganese. Carbon dioxide (CO_2) and hydrogen sulfide (H_2S) are produced during the aeration process. Air stripping has demonstrated efficacy in removing NH_3 from wastewater and eliminating volatile flavors and other substances from water. The Kimochi research group discovered that employing aerobic treatment with biowastes is an efficient method to decrease the emission of detrimental gaseous substances, including greenhouse gases and ammonia (Shannon et al., 2008).

1.5.3.3. Anaerobic Treatment

Anaerobic treatments are commonly used for deposition sludge supplemental treatment and for wastewater remediation with elevated amounts of biodegradable organic matter (BOD) surpassing 500 mg per liter. The agricultural and food processing industries generate a significant amount of organic wastewater that is rich in biodegradable materials. These types of wastewaters can be challenging and expensive to treat aerobically because of the high oxygen demand necessary to maintain the necessary aerobic environment. Anaerobic degradation, on the other hand, takes place in the absence of oxygen. While the anaerobic treatment method may need more time, there are numerous benefits when treating highly organic effluent (Rajeshwari et al., 2000). The advantages of this process include heightened purification levels, the capacity to manage substantial quantities of organic waste, minimal production of stable sludge, and the generation of methane as a byproduct. Anaerobic respiration is a complex process that involves several chemical steps. Organic materials break down into their fundamental components, ultimately leading to the generation of methane gas when there is no electron acceptor, such as oxygen (Narihiro & Sekiguchi, 2007).

Did You Know?

The capacity of biological treatment methods, including anaerobic digestion and activated sludge, to use microorganisms to break down organic contaminants into safe byproducts eliminates the need for energy-intensive procedures and chemical additives.

1.5.3.4. Activated Sludge

The activated sludge process entails the presence of a concentrated population of bacteria that are suspended in the wastewater, thereby establishing an environment rich in oxygen. In conditions of ample nutrients and oxygen, coupled with accelerated bacterial metabolism and respiration, the existing organic matter is either transformed into oxidized end-products or utilized for the proliferation of new microorganisms. The activated sludge process consists of five phases: the bioreactor phase, the activated sludge phase, the aeration and blending system, the sedimentation tank, and the returned sludge phase (Holenda et al., 2008). Wastewater treatment commonly utilizes biological processes, and one popular and cost-effective method is the activated sludge process (Figure 1.4). However, wastewater treatment plants (WWTPs) generate a substantial quantity of surplus sludge, which presents considerable difficulties in terms of expenses and environmental consequences. The treatment of the excessive sludge poses a significant challenge due to its elevated moisture content. The by-products of wastewater treatment plants are subjected to a series of procedures including dewatering, drying, and ultimately being converted into ashes through combustion. Some are used in agricultural settings as compost fertilizers. It is advisable to use dehydrated waste products from wastewater treatment plants in the pyrolysis process rather than incinerating them (Werle & Wilk, 2010). The aeration tank in the activated sludge process contains particulate matter that is held in suspension. The quantity of these particles in suspension is known as mixed liquor suspended solids (MLSS) and is measured in milligrams per liter. Mixed liquor refers to the amalgamation of untreated wastewater and activated sludge, which is present within an aeration tank (Marguti et al., 2018) (Figures 1.3 and 1.4).

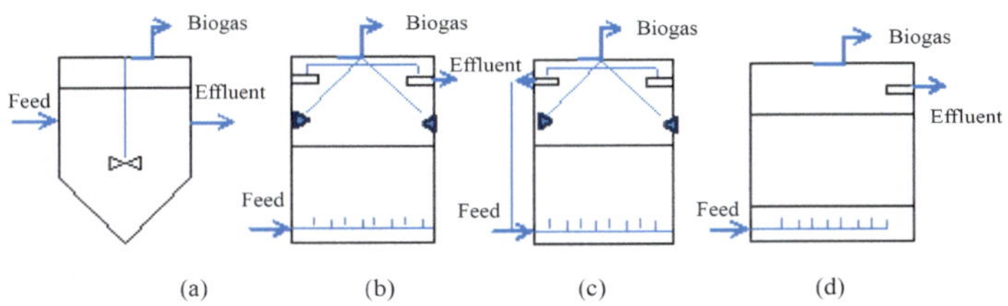

Figure 1.3. *Diagram of typical anaerobic digester reactor types.*

Source: Muaz Mohd Ziani, Creative Commons License.

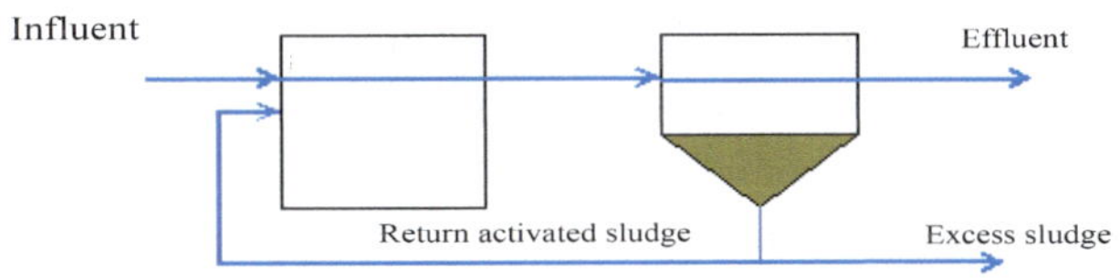

Figure 1.4. *Illustration of activated sludge process.*

Source: Muhammad Syazwan, Creative Commons License.

The MLSS is predominantly composed of microorganisms and suspended solids that exhibit limited susceptibility to degradation; and play a vital role in the activated sludge process by guaranteeing a sufficient quantity of active microorganisms to effectively degrade the organic pollutants. The term commonly used to describe this is the Food to Microorganism Ratio (F/M Ratio) or the food-to-mass ratio. When the ratio is maintained at the optimal level, the biomass develops the ability to effectively break down substantial quantities of food. As a result, this reduces the quantity of remaining food in the wastewater (Schwarz et al., 2006). To put it concisely, enhancing the decomposition of food will lead to decreased levels of biochemical oxygen demand (BOD) in the processed wastewater. It is essential to eliminate the Biochemical Oxygen Demand (BOD) from the MLSS to efficiently treat the wastewater and guarantee its suitability for different applications while upholding hygiene standards.

The wastewater treatment system receives untreated wastewater that has been exposed to a concentration of several 100 milligrams per liter of biochemical oxygen demand. Following treatment with MLSS and other techniques, the concentration of BOD in the wastewater decreases to below 2 mg L^{-1}, which is considered to be a safe level for utilization, as determined by Schwarz et al. (2006; Asada et al., 2010).

1.5.4. Chemical Treatments

During the initial phases of wastewater treatment technological progress, chemical treatment was employed before biological treatment. The biological treatment is currently carried out as a tertiary treatment and comes before the chemical treatment. Chemical treatment methods frequently employed encompass chemical precipitation, adsorption, neutralization, ion exchange, and disinfection (Saleh et al., 2020).

1.5.4.1. Chemical Precipitation

Eliminating dissolved inorganic material can be accomplished through various techniques, including pH adjustment, temperature alteration, or inducing precipitation as a solid. Precipitation can be eliminated by employing flotation, sedimentation, or other solid removal techniques. Although chemical precipitation is still utilized, it is strongly advised to contemplate phytoremediation as a substitute for this procedure. Bioremediation and phytoremediation are efficient techniques that help decrease the dependence on chemicals, in line with the increasing emphasis on environmental sustainability (Mbamba et al., 2015).

1.5.4.2. Neutralization

Neutralization involves maintaining the pH of the wastewater at a balanced level of around 7, regardless of whether it is alkaline or acidic. Ensuring the pH falls within the acceptable range necessitates the addition of a base due to insufficient alkalinity (Maree & Du Plessis, 1994). The most commonly used chemicals for pH adjustment are lime, calcium hydroxide, sodium hydroxide, and sodium carbonate. Ensuring the appropriate level of acidity is crucial, necessitating the inclusion of an acid to achieve the desired pH range. The pH of wastewater is often adjusted using commonly used chemicals such as sulfuric acid and carbonic acid (Filippidis et al., 2013).

1.5.4.3. Disinfection

The final stage in the tertiary treatment process involves disinfecting wastewater. The objective of disinfection is to eliminate or, at the very least, deactivate the harmful microorganisms. Effluent disinfection serves a dual purpose: safeguarding public health by eradicating or deactivating harmful pathogens like microbes, viruses, and protozoa, and ensuring compliance with wastewater effluent standards (Hand & Cusick, 2021).

The primary goal of disinfection is to safeguard the microbial integrity of wastewater. An effective disinfectant should demonstrate strong antibacterial properties, be cost-effective, easy to handle, and offer a dependable method for detecting any remaining traces. Extensive research has been conducted on the effectiveness of different chemical disinfection agents, such as chlorine, ultraviolet radiation, ozone, bromine, and chlorine dioxide (Haas et al., 1982).

1.5.4.4. Ion Exchange

Ion exchange is a reversible process in which a charged ion in a solution is exchanged with a similarly charged ion that is electrostatically bonded to a stationary solid particle. Ion exchange is a well-established technique in wastewater treatment, commonly employed for softening.

This process entails the substitution of polyvalent cations, such as calcium and magnesium, with sodium. Practically, wastewater is transferred into a stratum of resin (Jørgensen, 1979). The resin is formed by the polymerization of organic substances, resulting in a porous matrix. Sodium frequently replaces cations in the solution. Once the bed becomes fully saturated with exchanged ions, it will be promptly closed. To revitalize the bed, a solution of sodium with a high concentration must be circulated through it. Figure 1.5 shows a diagram of an organic cation-exchange bead (Verbych et al., 2005).

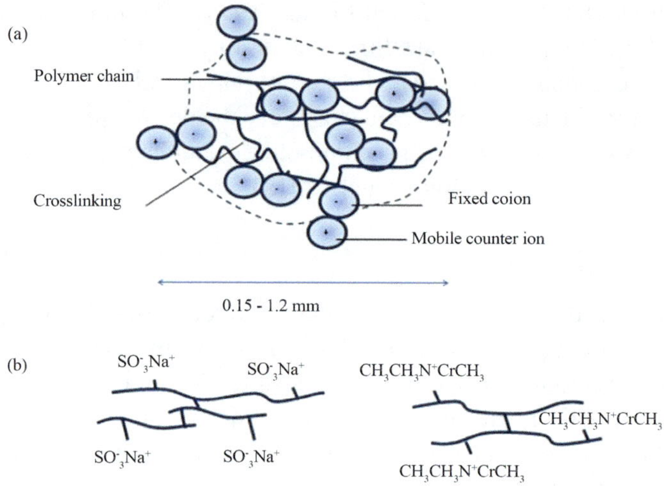

Figure 1.5. *Schematic of organic cation-exchange bead.*

Source: Nor Izzah, Creative Commons License.

1.5.5. Integrated Terminals

Advanced wastewater treatment is required when specific constituents in the wastewater cannot be effectively removed through secondary treatment alone. It is imperative to employ distinct treatment methods for phosphorus, nitrogen, extra suspended particles, refractory organic matter, heavy metals, and dissolved solids. Advanced treatment processes can be integrated with either primary or secondary treatment methods, such as the introduction of chemicals into primary clarifiers or aeration basins to eradicate phosphorus. Alternatively, they can serve as a substitute for secondary treatment, such as employing overland flow treatment for primary effluent. It is commonly known as integrated treatment. For example, when it comes to removing phosphorus, the usual secondary treatment only gets rid of around 1–2 mg/l of phosphorus. As a result, a substantial quantity of phosphorus is discharged into the final effluent, causing the excessive growth of algae and other aquatic plants in ponds and natural water streams, a process known as eutrophication (Rodrigues & de Aragão Umbuzeiro, 2011).

Typically, it exists as orthophosphate, polyphosphate, and phosphorus bonded to organic compounds. Microorganisms depend significantly on phosphorus for the synthesis, transportation, and storage of energy. Metcalf &

Eddy (2003) state that a considerable amount of phosphorus in the incoming wastewater is removed during the secondary treatment phase, usually falling between 10 and 30%. Currently, the method used to eliminate phosphate involves chemical precipitation, which can be quite expensive. Furthermore, it can result in a substantial augmentation in sludge volume, occasionally reaching up to 40%. To resolve the problem, the technique known as biological phosphate removal (BPR) is employed. This entails meticulously regulating the environmental parameters within the reactors to attain the intended result. Multiple treatment technologies exist for the elimination of phosphorus (de-Bashan & Bashan, 2004):

- **i.** **Physical:** Exploring filtration methods for removing particulate phosphorus and the potential of membrane technologies.
- **ii.** **Chemical:** Utilizing precipitation and physical-cum-chemical adsorption
- **iii.** **Biological:** Enhanced biological phosphorus removal (EBPR) and Assimilation (Song et al., 2020).

1.6. WATER QUALITY

Learning Objectives:

- Understand the assessment of parameters for drinking water quality.
- Comprehend WHO standards for quality of water assumed safe for human consumption.

Assessing the quality of water and wastewater can be done by considering the overall parameters used for evaluation. There are distinct differences in the parameters used to assess the quality of drinking water compared to wastewater. It is crucial to consider each parameter when evaluating the quality of both wastewater and water (Garner et al., 2021). When the concentration of certain parameters is high, it indicates a decline in the quality of water and wastewater. As an illustration, a significant presence of $CaCO_3$ in the drinking water can mimic the symptoms of kidney stones, leading to a classification of low water quality. Similar concerns arise with wastewater, as its discharge into streams can lead to water contamination and pose risks to both environment and living organisms (Keraita et al., 2003).

1.6.1. Parameters Assessment for Drinking Water Quality

Ensuring the safety of water for both livestock use and human consumption is of utmost importance. Thorough analysis and monitoring of water supplies are necessary to meet the required standards for bacterial and chemical composition. Most countries' local health departments are capable of conducting microbiological tests and chemical or mineral analyses. Testing procedures may vary depending on the type of contaminants being examined. Regrettably, it is not possible to obtain comprehensive information on all potential contaminants from a single water test (Petrović et al., 2003).

1.6.1.1. Mineral Analysis

An analysis of minerals can determine the inorganic composition present in water. A standard mineral analysis provides the concentration of mineral elements like magnesium, calcium, iron, manganese, copper, and zinc in parts per million. The analysis also presents the results of water acidity or pH and hardness, which are expressed in parts per million as well as grains per gallon (Baban, 1993). Additionally, it provides information on the levels of sulfates, nitrate, and other chemical substances. Significant quantities of minerals and other contaminants can potentially pose a risk to human health. As an example, the presence of nitrate in water can lead to health issues for both infants and ruminant animals. Similarly, sulfates can have a noticeable laxative effect on humans, while long-term consumption of arsenic has been linked to an increased risk of cancer. High concentrations of minerals can also impact the water's appearance and usability. High levels of magnesium and calcium contribute to the presence of hard water. Red stains can be left on plumbing fixtures, equipment, and laundry due to the presence of iron. Water can appear murky or hazy due to suspended silt, while an unpleasant taste or odor may be caused by dissolved gases (Pantelić et al., 2019).

1.6.1.2. Microbial Tests

Microbiological testing is used to assess the potential presence of harmful microorganisms in water, which can pose a risk to public health. Testing drinking water for potential pathogens can be a complex, time-consuming, and costly process. On the other hand, water is typically examined for indicator microorganisms like total coliform and *E. coli* bacteria. This is because the presence of these microorganisms in water suggests that the well and the surroundings could potentially harbor other harmful microorganisms (Conn et al., 2012). Therefore, if a water test shows the presence of total coliform or both total coliform and *E. coli,* it suggests the potential presence of harmful microorganisms that can cause diseases. Coliform bacteria are a diverse group of microorganisms. Many coliform bacteria are harmless and can be found abundantly in soils, water bodies, and vegetation. The detection of this bacteria in tap water suggests the potential presence of disease-causing organisms in the water system.

E. coli is a unique form of coliform bacteria that is typically associated with fecal contamination. The presence of *E. coli* bacteria is quite common in the intestines of both warm-blooded animals and humans, and majority of these bacteria are harmless. Nevertheless, certain strains of *E. coli* can result in illness. The detection of *E. coli* in a sample of drinking water typically suggests recent fecal contamination from sewage or manure, which increases the likelihood of the presence of harmful pathogens (LeJeune et al., 2001).

1.6.1.3. WHO Standard (World Health Organization) for Drinking Water

The World Health Organization (WHO) formulates comprehensive international guidelines for water quality and its impact on human health. These guidelines serve as a fundamental basis for the establishment of regulations and standards across the globe (Kroehler, 2014). The Guidelines for Drinking Water Quality (GDWQ) have the objective of safeguarding public health by facilitating the establishment of regulations and standards that are tailored to the unique conditions of each locality. They also encourage the implementation of proactive risk management techniques that address every link in the water supply chain, from the catchment area to the end user. The Water Safety Plans approach, referred to as a comprehensive strategy, has been acknowledged as a highly effective method for guaranteeing the safety of drinking water (Channar et al., 2014).

1.7. LIMITATION OF WATER AND WASTEWATER TREATMENT

Learning Objectives:

- Comprehend the limitations associated with treatment of wastewater.

Water and wastewater play crucial roles in sustaining life for both humans and all living creatures on the planet. Clean water is essential for the survival of all organisms, while waste is an inevitable byproduct of their existence. If the waste is not managed effectively, it can be discharged and added to the wastewater (Ahmad, 2022). Additionally, untreated wastewater can negatively affect groundwater and surface water resources. The financial component is one of the biggest problems in the field of wastewater and water treatment. Some treatment methods require significant energy consumption and may not be financially viable (Ziembowicz & Kida, 2022). Therefore, it is important to consider a range of treatment methods when choosing the most suitable approach for treatment. An excellent illustration is the treatment of small and regular domestic wastewater. Given the circumstances, it may be more appropriate and economical to consider biological treatment instead of more advanced methods like membrane filtration. In addition, it is important to take into account the costs associated with operation and maintenance. As an illustration, the process of coagulation and flocculation necessitates a significant amount of chemicals, resulting in substantial operational expenses. Regular maintenance is essential to ensure the optimal performance and longevity of mechanical machines, reactors, and tanks (Zieliński et al., 2022).

If these aspects are not taken into consideration, there may be treatment plant accidents or wastewater discharge that does not adhere to regulations. Ensuring safety is a challenge in the treatment of water and wastewater. It is vital to put worker, animal, and environmental safety first while handling toxic wastewater (Garrido et al., 2013). When considering the use of specific microorganisms in a treatment plant, it is essential to thoroughly assess the potential biohazard associated with them. When considering the ecosystem, it is crucial to ensure that the treatment methods are both efficient and sustainable. It is important to consider the potential impact of environmental factors, such as weather and natural disasters, on the treatment plant and its surrounding areas. For instance, a treatment plant located in an area with frequent heavy rainfall must assess the precipitation levels and tailor the design of the plant accordingly (Kontas et al., 2004).

SUMMARY

- This chapter highlights the significant consequences of polluted water sources on the transmission of waterborne illnesses. It also explores the vital importance of water as a basic resource for life and supporting different human activities. An extensive analysis of the various sources of wastewater is conducted, covering both residential and industrial origins.
- This chapter also explores the various components found in wastewater, such as industrial effluents, sewage, and agricultural runoff. Each of these components brings its own set of challenges when it comes to treatment and management.
- This chapter also emphasizes the historical development of wastewater treatment practices. This comprehensive review offers valuable insights into the gradual improvement and fine-tuning of wastewater treatment technologies throughout history, spurred by advancements in engineering, science, and public health awareness.
- The chapter organizes water and wastewater treatment techniques into different types, including chemical, physical, biological, and Advanced Oxidation Processes (AOPs). This organized categorization offers readers a thorough structure to grasp the wide range of treatment choices for tackling various water pollution challenges.
- This chapter also addresses the inherent limitations that are associated with wastewater treatment processes.

REVIEW QUESTIONS

1. Explain the significance of water quality in relation to public health. How does wastewater treatment contribute to the prevention of waterborne diseases?
2. Discuss the various sources of wastewater. How do domestic and industrial wastewater differ in terms of composition and treatment requirements?
3. Trace the historical development of wastewater treatment practices. What were some of the earliest methods employed for treating wastewater, and how have they evolved over time?
4. Compare and contrast the different types of water and wastewater treatment techniques discussed in the chapter.

5. Describe the operational principles and applications of Advanced Oxidation Processes (AOPs) in wastewater treatment.

MULTIPLE CHOICE QUESTIONS

1. **Which of the following best describes the relationship between water quality and public health?**
   a. Water quality has no impact on public health.
   b. Water quality is crucial for maintaining public health.
   c. Water quality only affects certain populations.
   d. Water quality is irrelevant to public health concerns.
2. **What is the primary role of wastewater treatment in preventing waterborne diseases?**
   a. Enhancing water aesthetics
   b. Reducing water availability
   c. Protecting public health
   d. Promoting industrial growth
3. **What are the main sources of wastewater?**
   a. Agricultural runoff only
   b. Industrial effluents only
   c. Domestic sewage only
   d. Both domestic and industrial sources
4. **What are the categories of water and wastewater treatment techniques mentioned in the chapter?**
   a. Physical, mechanical, biological
   b. Chemical, biological, thermal
   c. Physical, chemical, biological, AOPs
   d. Mechanical, thermal, AOPs, electrochemical
5. **What is the purpose of Advanced Oxidation Processes (AOPs) in wastewater treatment?**
   a. To add nutrients to wastewater
   b. To remove only physical contaminants
   c. To remove organic and inorganic contaminants
   d. To reduce water pH

Answers to Multiple Questions

1. (b); 2. (c); 3. (d); 4. (c); 5. (c)

REFERENCES

1. Ahmad, I. Z. (2022). The usage of cyanobacteria in wastewater treatment: Prospects and limitations. *Letters in Applied Microbiology, 75*(4), 718–730. https://doi.org/10.1111/lam.13734.
2. Asada, L. N., Sundefeld, G. C. Jr., Alvarez, C. R., Filho, S. S. F., & Piveli, R. P. (2010). Water treatment plant sludge discharge to wastewater treatment works: Effects on the operation of up flow anaerobic sludge blanket reactor and activated sludge systems. *Water Environment Research, 82*(5), 392–400. https://doi.org/10.2175/106143009X12589821300896.
3. Ashbolt, N. J. (2004). Microbial contamination of drinking water and disease outcomes in developing regions. *Toxicology, 198*(1–3), 229–238. https://doi.org/10.1016/j.tox.2004.01.020.
4. Augustos de Lemos Chernicharo, C., & Von Sperling, M. (2005). *Biological Wastewater Treatment in Warm Climate Regions* (Vol. 2, pp. 857–860). IWA Publishing.
5. Baban, S. M. (1993). Detecting water quality parameters in the Norfolk Broads, UK, using Landsat imagery. *International Journal of Remote Sensing, 14*(7), 1247–1267. https://doi.org/10.1080/01431169308954112.
6. Babu, D. S., Srivastava, V., Nidheesh, P. V., & Kumar, M. S. (2019). Detoxification of water and wastewater by advanced oxidation processes. *Science of the Total Environment, 696*, 133961. https://doi.org/10.1016/j.scitotenv.2019.133961.
7. Burger, R. A., Brady, M. F., Bookman, M. A., Fleming, G. F., Monk, B. J., Huang, H., & Liang, S. X. (2011). Incorporation of bevacizumab in the primary treatment of ovarian cancer. *New England Journal of Medicine, 365*(26), 2473–2483. https://doi.org/10.1056/NEJMoa1104391.
8. Channar, A. G., Rind, A. M., Mastoi, G. M., Almani, K. F., Lashari, K. H., Qurishi, M. A., & Mahar, N. (2014). Comparative study of water quality of Manchhar Lake with drinking water quality standard of World Health Organization. *American Journal of Environmental Protection, 3*(2), 68–72. https://doi.org/10.12691/ajep-3-2-2.
9. Conn, K. E., Habteselassie, M. Y., Blackwood, A. D., & Noble, R. T. (2012). Microbial water quality before and after the repair of a failing onsite wastewater treatment system adjacent to coastal waters. *Journal*

of Applied Microbiology, 112(1), 214–224. https://doi.org/10.1111/j.1365-2672.2011.05248.x.

10. De-Bashan, L. E., & Bashan, Y. (2004). Recent advances in removing phosphorus from wastewater and its future use as fertilizer (1997–2003). *Water Research, 38*(19), 4222–4246. https://doi.org/10.1016/j.watres.2004.07.014.
11. Deng, Y., & Zhao, R. (2015). Advanced oxidation processes (AOPs) in wastewater treatment. *Current Pollution Reports, 1*, 167–176. https://doi.org/10.1007/s40726-015-0015-7.
12. Filippidis, A., Godelitsas, A., Kantiranis, N., Gamaletsos, P., Tzamos, E., & Filippidis, S. (2013). Neutralization of sludge and purification of wastewater from Sindos industrial area of Thessaloniki (Greece) using natural zeolite. *Bulletin of the Geological Society of Greece, 47*(2), 920–926. https://doi.org/10.12681/bgsg.13372.
13. Finger, R. E., & Parrick, J. (1980). Optimization of grit removal at a wastewater treatment plant. *Journal (Water Pollution Control Federation), 52*(3), 2106–2116. https://doi.org/10.2307/25037956.
14. Galiani, S., Gertler, P., & Schargrodsky, E. (2005). Water for life: The impact of the privatization of water services on child mortality. *Journal of Political Economy, 113*(1), 83–120. https://doi.org/10.1086/426041.
15. Garner, E., Davis, B. C., Milligan, E., Blair, M. F., Keenum, I., Maile-Moskowitz, A., & Pruden, A. (2021). Next generation sequencing approaches to evaluate water and wastewater quality. *Water Research, 194*, 116907. https://doi.org/10.1016/j.watres.2021.116907.
16. Garrido, J. M., Fdz-Polanco, M., & Fdz-Polanco, F. (2013). Working with energy and mass balances: A conceptual framework to understand the limits of municipal wastewater treatment. *Water Science and Technology, 67*(10), 2294–2301. https://doi.org/10.2166/wst.2013.165.
17. Gedda, G., Balakrishnan, K., Devi, R. U., Shah, K. J., Gandhi, V., Gandh, V., & Shah, K. (2021). Introduction to conventional wastewater treatment technologies: Limitations and recent advances. *Materials Research Foundations, 91*, 1–36. https://doi.org/10.4028/www.scientific.net/MRF.91.1.
18. Gogate, P. R., & Pandit, A. B. (2004). Sono photocatalytic reactors for wastewater treatment: A critical review. *AIChE Journal, 50*(5), 1051–1079. https://doi.org/10.1002/aic.10152.
19. Haas, A. L., Warms, J. V., Hershko, A., & Rose, I. A. (1982). Ubiquitin-activating enzyme. Mechanism and role in protein-ubiquitin conjugation. *Journal of Biological Chemistry, 257*(5), 2543–2548. https://doi.org/10.1016/S0021-9258(18)51781-2.

20. Hand, S., & Cusick, R. D. (2021). Electrochemical disinfection in water and wastewater treatment: Identifying impacts of water quality and operating conditions on performance. *Environmental Science & Technology, 55*(6), 3470–3482. https://doi.org/10.1021/acs.est.0c07610.

21. Holenda, B., Domokos, E., Redey, A., & Fazakas, J. (2008). Dissolved oxygen control of the activated sludge wastewater treatment process using model predictive control. *Computers & Chemical Engineering, 32*(6), 1270–1278. https://doi.org/10.1016/j.compchemeng.2007.09.005.

22. Jørgensen, S. E. (1979). Examination of the applicability of cellulose ion exchangers for water and wastewater treatment. *Water Research, 13*(12), 1239–1247. https://doi.org/10.1016/0043-1354(79)90141-0.

23. Kataki, S., Chatterjee, S., Vairale, M. G., Dwivedi, S. K., & Gupta, D. K. (2021). Constructed wetland, an eco-technology for wastewater treatment: A review on types of wastewaters treated and components of the technology (macrophyte, biofilm and substrate). *Journal of Environmental Management, 283*, 111986. https://doi.org/10.1016/j.jenvman.2021.111986.

24. Keraita, B., Drechsel, P., & Amoah, P. (2003). Influence of urban wastewater on stream water quality and agriculture in and around Kumasi, Ghana. *Environment and Urbanization, 15*(2), 171–178. https://doi.org/10.1177/095624780301500212.

25. Kontas, A., Kucuksezgin, F., Altay, O., & Uluturhan, E. S. İ. N. (2004). Monitoring of eutrophication and nutrient limitation in the Izmir Bay (Turkey) before and after wastewater treatment plant. *Environment International, 29*(8), 1057–1062. https://doi.org/10.1016/j.envint.2003.11.004.

26. Kroehler, C. J. (2014). Potable water quality standards and regulations: A historical and world overview. In *Potable Water: Emerging Global Problems and Solutions* (pp. 1–36).

27. Kroiss, H., Matsché, N., & Krampe, J. (2021). 50 years of design and operation of large wastewater treatment plant conferences: A history of innovation and development. *Water Science and Technology, 84*(2), 263–273. https://doi.org/10.2166/wst.2021.039.

28. Lee, D. E., Ayoub, N., & Agrawal, D. K. (2016). Mesenchymal stem cells and cutaneous wound healing: Novel methods to increase cell delivery and therapeutic efficacy. *Stem Cell Research & Therapy, 7*, 1–8. https://doi.org/10.1186/s13287-016-0325-6.

29. LeJeune, J. T., Besser, T. E., & Hancock, D. D. (2001). Cattle water troughs as reservoirs of *Escherichia coli* O157. *Applied and Environmental*

Microbiology, 67(7), 3053–3057. https://doi.org/10.1128/AEM.67.7.3053-3057.2001.

30. Mahamuni, N. N., & Adewuyi, Y. G. (2010). Advanced oxidation processes (AOPs) involving ultrasound for wastewater treatment: A review with emphasis on cost estimation. *Ultrasonics Sonochemistry, 17*(6), 990–1003. https://doi.org/10.1016/j.ultsonch.2010.05.008.
31. Mahgoub, M. E. S. M., van der Steen, N. P., Abu-Zeid, K., & Vairavamoorthy, K. (2010). Towards sustainability in urban water: A life cycle analysis of the urban water system of Alexandria City, Egypt. *Journal of Cleaner Production, 18*(10–11), 1100–1106. https://doi.org/10.1016/j.jclepro.2010.03.011.
32. Maree, J. P., & Du Plessis, P. (1994). Neutralization of acid mine water with calcium carbonate. *Water Science and Technology, 29*(9), 285–296. https://doi.org/10.2166/wst.1994.0521.
33. Marguti, A. L., Ferreira Filho, S. S., & Piveli, R. P. (2018). Full-scale effects of addition of sludge from water treatment stations into processes of sewage treatment by conventional activated sludge. *Journal of Environmental Management, 215*, 283–293. https://doi.org/10.1016/j.jenvman.2018.03.075.
34. Mbamba, C. K., Batstone, D. J., Flores-Alsina, X., & Tait, S. (2015). A generalized chemical precipitation modelling approach in wastewater treatment applied to calcite. *Water Research, 68*, 342–353. https://doi.org/10.1016/j.watres.2014.10.058.
35. Miklos, D. B., Remy, C., Jekel, M., Linden, K. G., Drewes, J. E., & Hübner, U. (2018). Evaluation of advanced oxidation processes for water and wastewater treatment: A critical review. *Water Research, 139*, 118–131. https://doi.org/10.1016/j.watres.2018.03.031.
36. Mohajerani, M., Mehrvar, M., & Ein-Mozaffari, F. (2009). An overview of the integration of advanced oxidation technologies and other processes for water and wastewater treatment. *International Journal of Engineering, 3*(2), 120–146. https://doi.org/10.3923/ije.2009.120.146.
37. Narihiro, T., & Sekiguchi, Y. (2007). Microbial communities in anaerobic digestion processes for waste and wastewater treatment: A microbiological update. *Current Opinion in Biotechnology, 18*(3), 273–278. https://doi.org/10.1016/j.copbio.2007.04.002.
38. Nichols, G., Lane, C., Asgari, N., Verlander, N. Q., & Charlett, A. (2009). Rainfall and outbreaks of drinking water related disease in England and Wales. *Journal of Water and Health, 7*(1), 1–8. https://doi.org/10.2166/wh.2008.013.

39. Noor, A., Kutty, S. R. M., Isa, M. H., Farooqi, I. H., Affam, A. C., Birniwa, A. H., & Jagaba, A. H. (2023). Treatment innovation using biological methods in combination with physical treatment methods. In *The Treatment of Pharmaceutical Wastewater* (Vol. 3, pp. 217–245). Elsevier. https://doi.org/10.1016/B978-0-12-824273-7.00011-6.

40. Oller, I., Malato, S., & Sánchez-Pérez, J. A. (2011). Combination of advanced oxidation processes and biological treatments for wastewater decontamination—A review. *Science of the Total Environment, 409*(20), 4141–4166. https://doi.org/10.1016/j.scitotenv.2011.07.021.

41. Pantelić, N. Đ., Jaćimović, S., Štrbački, J., Milovanović, D. B., Dojčinović, B. P., & Kostić, A. Ž. (2019). Assessment of spa mineral water quality from Vrnjačka Banja, Serbia: Geochemical, bacteriological, and health risk aspects. *Environmental Monitoring and Assessment, 191*, 1–16. https://doi.org/10.1007/s10661-019-7468-1.

42. Parker, D. S. (2011). Introduction of new process technology into the wastewater treatment sector. *Water Environment Research, 83*(6), 483–497. https://doi.org/10.2175/106143011X13023873136922.

43. Petrie, B. (2021). A review of combined sewer overflows as a source of wastewater-derived emerging contaminants in the environment and their management. *Environmental Science and Pollution Research, 28*(25), 32095–32110. https://doi.org/10.1007/s11356-021-13437-8.

44. Petrović, M., Gonzalez, S., & Barceló, D. (2003). Analysis and removal of emerging contaminants in wastewater and drinking water. *TrAC Trends in Analytical Chemistry, 22*(10), 685–696. https://doi.org/10.1016/S0165-9936(03)01011-7.

45. Rajeshwari, K. V., Balakrishnan, M., Kansal, A., Lata, K., & Kishore, V. V. N. (2000). State-of-the-art of anaerobic digestion technology for industrial wastewater treatment. *Renewable and Sustainable Energy Reviews, 4*(2), 135–156. https://doi.org/10.1016/S1364-0321(99)00026-6.

46. Rezai, B., & Allahkarami, E. (2021). Wastewater treatment processes—Techniques, technologies, challenges faced, and alternative solutions. In *Soft Computing Techniques in Solid Waste and Wastewater Management* (Vol. 3, pp. 35–53). Elsevier. https://doi.org/10.1016/B978-0-12-821388-1.00004-1.

47. Rodrigues, E. S., & de Aragão Umbuzeiro, G. (2011). Integrating toxicity testing in the wastewater management of chemical storage terminals: A proposal based on a ten-year study. *Journal of Hazardous Materials, 186*(2–3), 1909–1915. https://doi.org/10.1016/j.jhazmat.2010.12.030.

48. Saleh, I. A., Zouari, N., & Al-Ghouti, M. A. (2020). Removal of pesticides

from water and wastewater: Chemical, physical and biological treatment approaches. *Environmental Technology & Innovation, 19*, 101026. https://doi.org/10.1016/j.eti.2020.101026.

49. Samer, M. (2015). Biological and chemical wastewater treatment processes. *Wastewater Treatment Engineering, 150*(10.5772), 61250. https://doi.org/10.5772/61250.
50. Schwarz, J. P., Gao, R. S., Fahey, D. W., Thomson, D. S., Watts, L. A., Wilson, J. C., & Aikin, K. C. (2006). Single-particle measurements of midlatitude black carbon and light-scattering aerosols from the boundary layer to the lower stratosphere. *Journal of Geophysical Research: Atmospheres, 111*(16), 852–855. https://doi.org/10.1029/2005JD006379.
51. Shannon, M. A., Bohn, P. W., Elimelech, M., Georgiadis, J. G., Marinas, B. J., & Mayes, A. M. (2008). Science and technology for water purification in the coming decades. *Nature, 452*(7185), 301–310. https://doi.org/10.1038/nature06599.
52. Shete, B. S., & Shinkar, N. P. (2013). Dairy industry wastewater sources, characteristics & its effects on environment. *International Journal of Current Engineering and Technology, 3*(5), 1611–1615. https://doi.org/10.14741/ijcet/3.5.5.
53. Sinclair, R. G., Jones, E. L., & Gerba, C. P. (2009). Viruses in recreational water-borne disease outbreaks: A review. *Journal of Applied Microbiology, 107*(6), 1769–1780. https://doi.org/10.1111/j.1365-2672.2009.04357.x.
54. Song, X., Liu, R., Yu, Q., Sheng, X., & Chen, L. (2020). Management mode construction for operation and supervision of rural sewage treatment facilities: Towards the information-to-intelligence strategy. *Bioresource Technology Reports, 11*, 100481. https://doi.org/10.1016/j.biteb.2020.100481.
55. Sontrop, J. M., Dixon, S. N., Garg, A. X., Buendia-Jimenez, I., Dohein, O., Huang, S. H. S., & Clark, W. F. (2013). Association between water intake, chronic kidney disease, and cardiovascular disease: A cross-sectional analysis of NHANES data. *American Journal of Nephrology, 37*(5), 434–442. https://doi.org/10.1159/000350917.
56. Spielvogel, D., Etz, C. D., Silovitz, D., Lansman, S. L., & Griepp, R. B. (2007). Aortic arch replacement with a trifurcated graft. *The Annals of Thoracic Surgery, 83*(2), S791–S795. https://doi.org/10.1016/j.athoracsur.2006.10.050.
57. Sturm, M., Goldstein, M. A., & Parr, C. (2017). Water and life from snow: A trillion-dollar science question. *Water Resources Research, 53*(5), 3534–3544. https://doi.org/10.1002/2016WR020036.

58. Verbych, S., Hilal, N., Sorokin, G., & Leaper, M. (2005). Ion exchange extraction of heavy metal ions from wastewater. *Separation Science and Technology, 39*(9), 2031–2040. https://doi.org/10.1081/SS-200063527.
59. Werle, S., & Wilk, R. K. (2010). A review of methods for the thermal utilization of sewage sludge: The Polish perspective. *Renewable Energy, 35*(9), 1914–1919. https://doi.org/10.1016/j.renene.2009.12.051.
60. Xu, A., Wu, Y. H., Chen, Z., Wu, G., Wu, Q., Ling, F., & Hu, H. Y. (2020). Towards the new era of wastewater treatment of China: Development history, current status, and future directions. *Water Cycle, 1*, 80–87. https://doi.org/10.1016/j.wacy.2020.05.002.
61. Zieliński, M., Kazimierowicz, J., & Dębowski, M. (2022). Advantages and limitations of anaerobic wastewater treatment—Technological basics, development directions, and technological innovations. *Energies, 16*(1), 83. https://doi.org/10.3390/en16010083.
62. Ziembowicz, S., & Kida, M. (2022). Limitations and future directions of application of the Fenton-like process in micropollutants degradation in water and wastewater treatment: A critical review. *Chemosphere, 296*, 134041. https://doi.org/10.1016/j.chemosphere.2022.134041.

CHAPTER 2

FUNDAMENTALS OF ADVANCED OXIDATION PROCESSES (AOPS)

LEARNING OBJECTIVES

At the end of this chapter, readers will be able to:

1. Understand the concept of AOPs and their role in wastewater treatment.
2. Comprehend different types of AOPs including ozone-based processes, UV irradiation-based processes, electrochemical processes, and physical AOPs.
3. Understand the advantages and limitations associated with conventional AOPs.
4. Gain knowledge about future perspectives and research directions in the field of AOPs.

INTRODUCTORY EXAMPLE

Sol Chemicals, a leading manufacturer of specialty chemicals, operates a large-scale production facility with a commitment to environmental stewardship. However, the manufacturing processes generate wastewater containing a diverse array of organic pollutants, presenting a significant challenge for traditional treatment methods. In their pursuit of sustainable wastewater management, Sol Chemicals has embarked on a journey to explore Advanced Oxidation Processes (AOPs) as a potential solution.

Sol Chemicals wants to minimize their ecological footprint, ensure compliance with strict environmental standards, and remove all organic contaminants from their wastewater streams completely by utilizing AOPs. Through collaboration with research institutions and technology providers, Sol Chemicals is pioneering the adoption of AOPs tailored to their specific wastewater composition and treatment needs.

UNIT INTRODUCTION

Water is the prime natural resource on our planet. Without water, life cannot be possible beyond a few days and the lack of access to adequate water leads to the spread of many diseases. As we know that the ocean contains most of the water on the planet and is highly prone to getting polluted. Major sources of water contamination are agriculture, construction, industrialization, and marine dumping. The studies in last 10 years show that industrialization, tourism, and fishing occupation are mostly affecting surface water and pose too many diseases.

The ugly unregulated trend of releasing effluents in the water bodies comprising organic dyes, plastic, toxic chemicals, and agricultural waste destroying the quality of water. Presently, the convention method for treating water has drawbacks and challenges with satisfactory results. The adsorption process has the drawback that it generates a huge amount of toxic sludge, convert into another form of pollutant, and remain in the ecosystem.

With the increasing civilization, we have to focus on the techniques that are able to meet up the growing demand for clean water. Hence, advanced oxidation method for the treatment of water was developed. This method is one of the most efficient methods used for the oxidation of biological pollutants that are not conceived by the convention method due to their chemical stability. The mechanism of degradation involves the production of reactive and unstable products like ozone, hydroxyl, and hydrogen peroxide radicals that react with the pollutant into decomposed product and other inorganic salts. In the reaction, hydroxyl radical (•OH) undergoes electron transfer or abstraction of hydrogen and forms a double bond with the pollutant. Advanced oxidation technology involves the four different methods for producing •OH using different reacting system.

2.1. ADVANCED OXIDATION PROCESS (AOP)

Learning Objectives:

- Understand the concept of AOPs.
- Comprehend the stages involved in advanced oxidation process.

AOP is a process that targets to create hydroxyl radicals (OH·) to oxidize contaminants in water. One way to define oxidation is as the process of removing electrons, which results in an increased oxidation state. For example, the transformation of Fe^{2+} into Fe^{3+} is an example of oxidation. As far as organic compounds are concerned, the process entails the formation of compounds that contain oxygen, like carbonyl compounds, alcohols, and organic acids. According to Naddeo et al. (2011), the hydroxyl radical possesses several advantageous characteristics, including a high oxidation potential of 2.8 V, the ease with which it can be generated in natural water conditions, a short duration of effectiveness, and the fact that it is a relatively non-selective electrophilic oxidant. Deng & Zhao (2015) state that there are four main ways in which hydroxyl radicals break down organic contaminants. These processes include electron transfer, radical combination, radical addition, and hydrogen abstraction. The hydroxyl radical-AOPs can be divided into three distinct groups, according to general classification (Miklos et al., 2018):

- AOPs based on UV;
- AOPs based on ozone;
- AOPs based on Fenton.

Advanced oxidation involves these basic steps.

i. The first step involves the formation of strong oxidants like •OH, HO_2•, O_2•$^-$, etc.

ii. In the second step, these oxidants react with organic contaminants present in the wastewater converting them into biodegradable compounds.

iii. The last step is the oxidation of these biodegradable intermediates leading to complete mineralization in water, carbon dioxide, and inorganic salts.

In advanced oxidation processes that are based on ultraviolet light, the most important step is the excitation of the substrate, which results in the transfer of electrons from the substrate to the ground state of molecular oxygen (O_2). During the process of homolytic cleavage, free radicals are produced. These free radicals then encounter dissolved oxygen, which outcomes in the formation of peroxide radicals.

When several different components, including OH^-, H_2O, and O_2^-, interact with the surface of the catalyst, the valence band has the potential to acquire a positive charge. This, in turn, leads to the production of hydroxyl radicals through the action of negative electrons (Babu et al., 2019). When UV irradiation comes into contact with oxidants like H_2O_2 or O_3, it has the potential to generate additional (OH·).

Iron, a substance that is commonly employed as a catalyst, is utilized to stimulate the formation of hydroxyl radicals (OH·) in water through the activation of hydrogen peroxide. Because of this, the Fenton process is the name given to the AOP that involves the utilization of iron products. In accordance with the conventional Fenton radical mechanism, the Dark Fenton reaction takes place in an acidic environment.

First, iron can be found in either the Fe^{2+} or Fe^{3+} oxidation states. Both of these states are possible. Furthermore, the initial oxidation state of iron does not have any influence on the overall efficiency of the degradation process. Several different processes that are derived from the traditional Fenton process have been proposed, including the following: The electro-Fenton reaction, the Fenton-like reaction, and the photo-Fenton reaction are all reactions that have been described by Deng & Zhao (2015).

There is a significant oxidizing capacity in ozone, which is characterized by a high oxidation potential of 2.07 V. However, for organic compounds to be put through the process of direct oxidation of oxygen, they must be in a state that is ionized and dissociated, rather than in their stable form. Because it has a direct influence on the process by which hydroxide ions break down ozone, the pH level of water is an extremely important factor. Depending on the specific circumstances, the presence of additional oxidants or irradiation can significantly increase the concentration of OH. There are currently three different pathways that have been found for advanced oxidation processes that are based on ozone (Mahamuni & Adewuyi, 2010) (Figure 2.1).

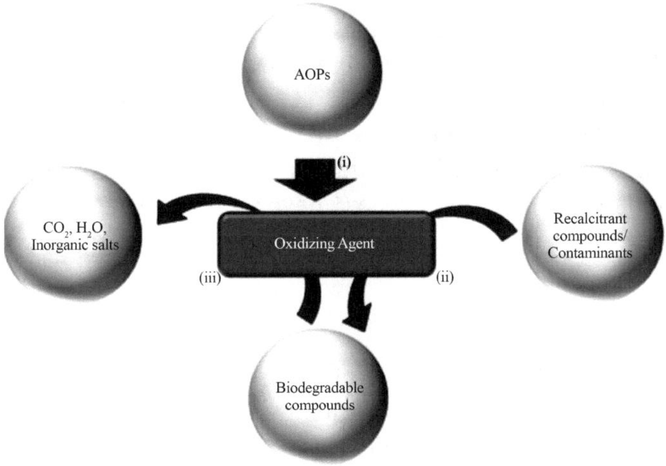

Figure 2.1. *Representation of the mechanism of advanced oxidation process.*

Source: Suresh Ameta, Creative Commons License.

2.2. OZONE-BASED AOPS

Learning Objectives:

- Understand the mechanism of ozone-based AOP.
- Comprehend different types such as ozonation, O3/H2O2 and O3/ catalysts.

Ozone is used as both an oxidizing agent and an insecticide in the process of water treatment. Ozone-based advanced oxidation processes (AOPs) primarily target electron-rich functional groups such as double bonds, amides, and aromatic compounds. This is because they generate hydroxyl radicals (•OH) in the aqueous solution. The reaction involved in these AOPs is a second-order reaction, which is somewhat slow with a rate constant of 70 M^{-1} s^{-1} (Kusic et al., 2006). Since O_3 treatment decolorizes bleaching water used in the paper, pulp, and textile sectors and successfully oxidizes a wide variety of contaminants, it is proven to be a more effective advanced oxidation approach. It is frequently utilized to treat pharmaceuticals in water by mixing O_3 with an oxidant (Liu et al., 2021). In an aqueous medium, the O_3 gas exhibits complete instability, leading to its direct attack on the substrate or the generation of •OH, which attacks the substrate. There are various techniques available for the production of O_3, including corona discharge, photocatalytic methods, and oxygen molecule conversion. UV radiation and corona discharge are the most ecologically friendly and cost-effective technologies. Temperature control is crucial in the corona discharge process because O_3 breakdown begins at high temperatures. As O_3 breaks down spontaneously at 35°C, it needs to have a cooling system installed. The following equations describe how oxygen gas splits into oxygen atoms, or reactive radicals, when it travels across the electrode and interacts with oxygen molecules to form ozone (Chávez et al., 2019).

$$O_2 + e^- \rightarrow 2O^-$$

$$O_2 + O \rightarrow O_3$$

The process of UV radiation involves the photo-dissociation of oxygen, which produces an unstable O_2 atom with a wavelength of 140–190 nm. Over 180 nm, however, the O_3 gas begins to break down. Therefore, due to the low yield of O_3 gas outputs, this approach is not commonly used in industrial locations (Wu et al., 2004) (Figure 2.2).

Did You Know?

Ozone was initially suggested for AOPs during the early 20th century. However, it wasn't until the latter half of the century that it gained significant interest due to its remarkable ability to degrade pollutants.

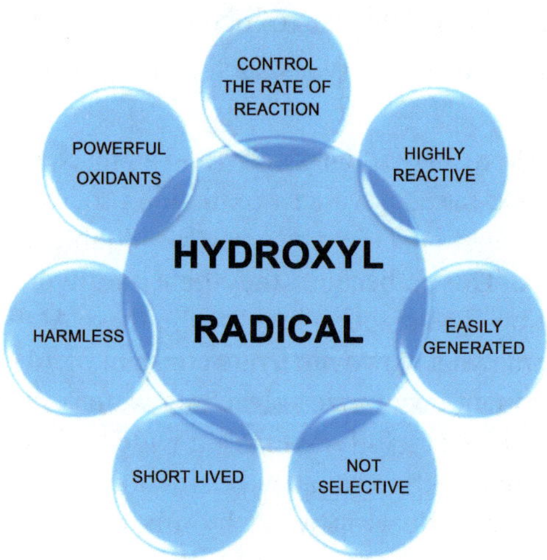

Figure 2.2. *Features of hydroxyl radical.*

Source: Pallavi Jain, Creative Commons License.

2.2.1. Ozonation at Elevated pH

There are two methods by which ozone can oxidize a material:

- The process involves an initial step of indirect oxidation, which is then followed by the generation of (•OH).
- Utilizing dissolved O_3 for direct oxidation.

The pH level of the medium plays a crucial role in determining the degradation of a compound. In most cases, if the pH is below 4, direct ozonation is the preferred method.

$$3O_3 + OH^- + H^+ \rightarrow 2OH + 4O_2$$

When the pH level is above 9, indirect oxidation emerges as the dominant process. As per the rule, the degradation rate in ozonation tends to increase as the pH level rises. Therefore, a higher pH level encourages the production of radicals, leading to indirect oxidation (Cortez et al., 2011).

$$O_3 + OH^- \rightarrow O_2 + HO_2^-$$

2.2.2. O_3/H_2O_2

A key component of this process is the reaction between O_3 and peroxide (H_2O_2), resulting in the formation of •OH precursors that serve as powerful oxidizing radicals. Essentially, the reaction between O_3 and H_2O_2 (in ionized form, hydroperoxide H_2O^-) resulted in the rapid generation of •OH radicals. This process is highly effective due to the enhanced decomposition rate of O_3, resulting in the generation of a substantial number of highly reactive •OH radicals.

$$H_2O_2 + 2O_3 \rightarrow 2^{\bullet}OH + 2O_3$$

$$H_2O_2 \leftrightarrow H^+ + H_2O^-$$

$$H_2O^- + O_3 \rightarrow {}^{\bullet}OH + {}^{\bullet}O_2 + O_2$$

The radicals are used to breakdown H_2O_2, with additional mechanisms occurring under optimal conditions, such as an H_2O_2/O_3 ratio of 0.5 and a pH of 7.7. This method is utilized to decrease the presence of micro-pollutants in industrial water, drinking water, and groundwater. Additionally, it is used to minimize the development of bromate (Guo et al., 2020).

2.2.3. O_3/Catalysts

Using catalysts in ozonation enhances the speed of ozone decomposition, resulting in the production of hydroxyl radicals. For the breakdown of organic contaminants, the homogeneous catalytic oxidation approach commonly uses d-block elements as catalysts.

Within this process, the presence of metal ions significantly impacts the rate of reaction, the effectiveness of ozone utilization, and the selectivity of ozone oxidation. There are two approaches that can be used to achieve homogenous catalytic ozonation:

- The production of free radicals by the breakdown of O_3 with the help of metal ions.
- Formation of complexes between organic molecules and catalysts through complex oxidation (Tanos et al., 2024).

$$Fe^{2+} + O_3 \rightarrow FeO^{2+} + O^{\bullet}$$

$$FeO^{2+} + H_2O \rightarrow Fe^{3+} + OH^{\bullet} + OH^{-}$$

$$FeO^{2+} + Fe^{2+} + 2H^{+} \rightarrow 2Fe^{3+} + H_2O$$

Organic molecules that are hard to biodegrade can be effectively ozoned by using solid catalysts in the heterogeneous catalytic ozonation process. Frequently used catalysts in the procedure include metal oxides like manganese oxide, iron oxide, and copper-iron oxide, as well as porous materials such as zeolite and activated carbon. The effectiveness of these catalytic methods depends on factors such as the pH of the solution, the specific pollutant being targeted, and the type of catalyst used. There are two steps involved in achieving heterogeneous catalytic ozonation: the hydroxyl method and the interfacial reaction method (Muruganandham et al., 2014).

2.3. PHYSICAL AOPS

Learning Objectives:

- Understand the mechanism of different types of physical AOPs including microwave, electron beam and plasma AOP.

2.3.1. Microwave (MW) AOP

For an efficient method of oxidizing water contaminants, the use of extremely energetic waves or radiation within the microwave area (300 MHz–300 GHz) has been explored. Microwaves are used in conjunction with oxidants (mostly

H_2O_2) and catalysts to help in the destruction of organic pollutants. These microwaves have the power to speed up reactions and cause internal molecule vibrations, which aid in the specific heating of different pollutants.

In addition, UV radiations can be generated by microwaves using an electrode-less discharge lamp in order to create united UV/MW reactors. Unfortunately, a significant portion of the microwave energy applied is converted into heat. Because of the low electrical productivity, it is necessary to use different types of cooling appliances to prevent the purified water from overheating (Shiying et al., 2009).

2.3.2. Electron Beam AOP

Since 1980, researchers have been studying the use of an electron beam source (0.01–10 meV) to generate ionizing radiation for wastewater treatment. Electrons can quickly enter the water, resulting in the creation of a species that is highly stimulated electronically. This species includes many ionic species and free radicals.

There is a direct proportionality between the maximal penetration strength of fast-moving electrons and the energy of the incident electrons, which is around 7 mm. The electron beam technique demonstrates a significant level of oxidizing power and minimal penetration through water and surrounding materials while remaining within the range of practical feasibility in terms of electrical capacity (Kim et al., 2012).

2.3.3. Plasma

Electrically liberating reactors in the aqueous phase are frequently used in wastewater using advanced oxidation processes. This process involves the application of a strong electric field within a liquid medium or between a gas phase and water.

This initiates a combination of physical and chemical processes. Besides rapidly oxidizing pollutants in water, revelation waves, a variety of active radicals, and UV radiations are generated upon discharge, which can promote additional oxidation (Chen et al., 2023).

Practice Problem

Describe the mechanism of photolysis in UV-based Advanced Oxidation Processes (AOPs) and its significance in pollutant degradation.

Solution to Practice Problem

Photolysis is a key mechanism in UV-based AOPs, where organic pollutants are degraded through direct photodecomposition induced by ultraviolet (UV) irradiation. When UV light of sufficient energy is absorbed by organic molecules in wastewater, it promotes the breaking of chemical bonds within these molecules, leading to the formation of radical species such as hydroxyl radicals (•OH). These highly reactive radicals then initiate chain reactions, attacking neighboring organic compounds and breaking them down into simpler, less toxic substances.

The significance of photolysis in UV-based AOPs lies in its ability to directly degrade organic pollutants without the need for additional chemical oxidants or catalysts. It offers a rapid and efficient method for pollutant removal, particularly for compounds that are susceptible to photodegradation under UV irradiation.

2.4. CATALYTIC AOPS

Learning Objectives:

- Understand the mechanism of different types of catalytic AOPs including fenton, photo-fenton and photocatalysis AOP.
- To discuss how hacker is a threat for the computer systems security.

2.4.1. Fenton AOP

The Fenton process is highly favored due to its exceptional capability to fully degrade pollutants, transforming them into harmless substances such as H_2O, CO_2, and various inorganic salts. When ferrous ions Fe^{2+} and H_2O_2 are combined under acidic conditions, it results in the generation of •OH, which is referred to as the Fenton reaction. To achieve the best catalytic activity of iron, it is crucial to maintain a pH of 3.0. This is because, at higher pH levels, there is a tendency for ferric oxyhydroxide to precipitate.

Fe^{3+} is reduced to Fe^{2+} by the excess injection of H_2O_2. Transition group metals can be used in place of iron oxide (FeO) to boost the trace organic compound elimination activity. To prevent iron precipitation, an acidic environment is maintained during the Fenton process. For this reason, an alternative to iron-free Fenton techniques has been looked at and summarized. The technique is more practical and cost-effective because of its cheap operating costs and simple residual iron separation using a magnet. The Fenton process has thus been established in several large-scale applications (Soon & Hameed, 2011).

2.4.2. Photo-Fenton AOP

The photo-Fenton process involves the combination of UV radiation and H_2O_2 with Fe^{2+} and Fe^{3+}, resulting in a higher generation of •OH compared to the Fenton technique or photolysis. This increase in •OH production leads to an enhanced rate of degradation of organic pollutants. In the photo-Fenton reaction, the concentration of Fe^{3+} increases in the solution. If all the Fe^{2+} ions are used up, the reaction will cease.

$$Fe(OH)^{2+} + h\nu \rightarrow Fe^{2+} + {}^{\bullet}OH$$

The photo-reduction of ferric ion Fe^{3+}, which takes place in the photo-Fenton reaction, is represented by the equation above as the regeneration of ferrous ions. This newly generated ferrous ion reacts with H_2O_2 to make •OH and ferric ions, and the cycle repeats itself in this way. Within the traditional visible and near-UV radiation zones, organic pollutants exhibit superior degradation through the Fenton reaction. The photo-Fenton approach significantly decomposed ethylene glycol, herbicides, anisole, and 4-chlorophenol, among other contaminants.

Direct photolysis of H_2O_2 is utilized to obtain •OH, which is useful in the breakdown of organic pollutants. The photo-Fenton process requires a pH of 3.0, soluble hydroxy-Fe^{3+} and higher photoactivity from Fe $(OH)^{2+}$ to function at its best. The photo-Fenton process produces superior results than the Fenton process. Sometimes UV radiation is substituted with sunlight as a cost-effective method. The rate at which pollutants degrade is slowed down by this method of using sunshine. Acidic conditions (pH 3) are preferred because carbonate and bicarbonate family members convert to carbonic acid at this pH and have a low rate of reactivity with •OH (Palaniandy et al., 2015).

2.4.3. Photocatalysis AOP

Because the photocatalytic process can handle a wide range of organic contaminants under normal temperature and pressure settings, it is attracting a lot of interest. This method involves exciting and transferring electrons from the valence band to the empty conduction band. The excitation of electrons occurs when the catalyst absorbs sufficient energy to overcome the band gap.

The photocatalytic process is generating significant interest due to its ability to address a diverse array of organic pollutants under normal temperature and pressure conditions. This method involves exciting and transferring electrons from the valence band to the empty conduction band. The excitation of electrons occurs when the catalyst absorbs sufficient energy to overcome the band gap (Rana et al., 2021) (Figure 2.3).

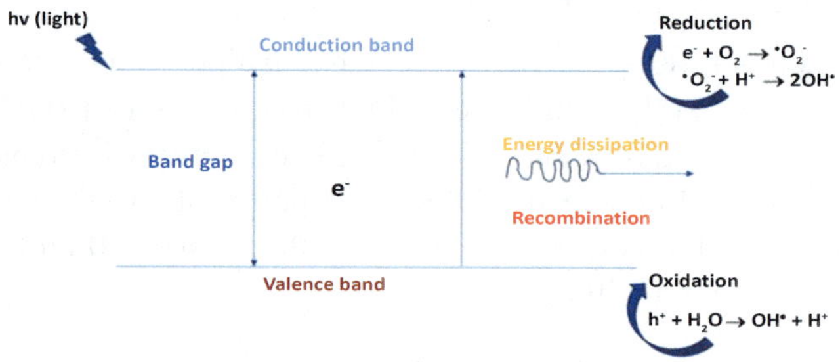

Figure 2.3. *Mechanism of photocatalysis AOP.*

Source: Prashant Singh, Creative Commons License.

Photocatalysts such as TiO_2 and ZnO_2 are commonly utilized due to their chemical stability and strong photosensitivity. However, these catalysts must possess effective reducing and oxidizing power to function optimally. Nevertheless, the high bandgap (3.2 eV) of these catalysts makes them less suitable for the intended application. To enhance their efficiency and stability, one can consider reducing the regeneration of h^+/e^- pairs through the use of TiO_2 and ZnO_2 doping. For doping, the electron of the conduction band or holes in the valance band is restricted to detect sites, resulting in a decrease in h^+/e^- pairs and enhancing interfacial charge transfer. Subsequently, it decreases the band gap and facilitates the absorption that takes place in the visible light spectrum (Xiao et al., 2020).

$$TiO_2 + h\nu \rightarrow TiO_2(e^-) + TiO_2(h^+)$$

$$TiO_2(e^-) + O_2 \rightarrow TiO_2 + O_2^-$$

$$TiO_2(h^+) + H_2O \rightarrow TiO_2 + {}^\bullet OH + H$$

$$TiO_2(h^+) + OH^- \rightarrow TiO_2 + {}^\bullet OH$$

Did You Know?

Certain advanced oxidation processes, like photocatalysis, can be executed with immobilized catalysts. This enables the implementation of continuous treatment procedures and facilitates the effortless separation of catalysts from the treated water.

2.5. ELECTROCHEMICAL AOPS

Learning Objectives:

- Understand the concept of electrochemical oxidation process.
- Comprehend the mechanism of different types of electrochemical AOPs including BDD, lead oxide and doped SnO2 electrodes.

The electrochemical oxidation process involves the generation of (•OH) to oxidize pollutants, without the production of any secondary sludge. The process of breaking down organic matter involves the interaction of a highly reactive radical with the substance, resulting in addition, substitution, or electron transfer reactions. These reactions effectively decompose the pollutant into organic matter. The production of (•OH) remains unaffected by changes in pH and temperature. Nevertheless, the effectiveness of water treatment is influenced by factors such as pH, temperature, and the presence of carbonate ions (Guelfi et al., 2018).

2.5.1. Boron-Doped Diamond Electrodes

In electrochemical processes, the boron-doped diamond electrode has emerged as a highly favored and extensively studied option. The most common process used to create these electrodes is chemical vapor deposition (CVD), which is favored because of how economical it is. Because of boron's low-point charge carrier activation energy of 0.37 eV, it was selected as a dopant for the diamond electrode. A p-type semiconductor is produced in the diamond lattice when boron atoms replace carbon atoms. The additional electron in this semiconductor is used for bonding, which creates holes.

At low levels of doping, around ($\sim 10^{17}$ atoms cm^{-3}) diamond exhibits semiconductor properties with conduction occurring through the hopping method. Due to low-energy impurity bands, diamond displays a semi-metallic property at high doping levels ($\sim 10^{20}$ atoms cm^{-3}). Because of this, electrons can conduct, leading to metallic resistivities that are usually less than 0.1 Ω cm (Huang et al., 2021).

2.5.2. Doped TiO_2 Electrode

Efficient electrode materials have emerged for water processing, specifically the conductive Magneli phase suboxide of doped TiO_2 and TiO_2. The electrical conductivity of stoichiometric TiO_2 is approximately $\sim 10^{-9}$–1 cm^{-1}, indicating its insulating properties. When exposed to temperatures above 900°C in an H_2 atmosphere, the electronic properties of TiO_2 can undergo significant changes. This can be achieved by creating O_2 deficiencies in the lattice structure or by introducing group five elements as dopants. These alterations lead to the formation of Ti^{4+} to Ti^{3+} and give rise to n-type semiconductor behavior (Zhou et al., 2023).

2.5.3. Lead Oxide Electrode

Previous research has utilized packed bed reactors with oxidized Pb pellets to investigate compound oxidation at PbO_2. Through extensive research on PbO_2 and doped PbO_2 anodes with different substrates such as Ta and Ti, numerous studies have revealed the generation of hydroxyl radicals on PbO_2 electrodes. However, the formation of •OH at the PbO_2 electrode has not been thoroughly investigated. The research revolves around the investigation of PbO_2 electrodes, specifically in relation to lead-acid batteries. The study also examines the generation of (•OH) within a hydrated PbO_2 gel layer that forms on the outer surface of the electrode. Within this approach, it is assumed that a balance is present between the hydrated gel layer of PbO_2 and the PbO_2 crystal. A gel layer forms a photon and an electrically conductive linear polymer chain (Zhuo et al., 2016).

2.5.4. Doped SnO_2 Electrode

To achieve the highest level of conductivity in the SnO_2 electrode, it is necessary to introduce doping, which enhances its performance as an efficient electrode for advanced oxidation processes. A dopant is utilized to enhance the electrode conductivity and enable the potential for O_2 evolution in comparison to SHE. Here, Sb is the chosen dopant. However, the use of Sb as a dopant can be detrimental due to its toxic nature. That is why researchers are exploring alternative methods (B, F, P, Ar). After extensive research and analysis, researchers have determined that the utilization of doped SnO_2 electrodes is primarily limited to laboratory settings, rather than being implemented on a large scale or in commercial applications, due to its limited lifespan.

Two theories have been proposed to explain the deactivation of these electrodes. The first theory suggests that a nonconductive $Sn (OH)^{2}$-layer forms on the outer surface of the anode. The second theory proposes that delamination occurs in the SnO_2 film due to the formation of a layer on the underlying Ti substrate. To enhance the performance of the doped SnO_2 electrode, it is necessary to reduce the layering of the Ti support. This can be achieved by inserting an IrO_2 interlayer between the Ti and SnO_2–Sb_2O_5 coating (Yang et al., 2022).

2.6. ADVANTAGES AND LIMITATIONS OF CONVENTIONAL AOPS

Learning Objectives:

- Comprehend the benefits and limitations of AOPs.

AOPs have demonstrated exceptional chemical stability, making them highly effective in eliminating contaminants that cannot be addressed by traditional methods. However, there are certain limitations that are addressed in Table 2.1 provided (Ma et al., 2021).

Table 2.1. *Advantages and Limitations of Conventional Advanced Oxidation Processes*

AOPs	Advantages	Limitations
O_3/UV	Better outcome with high efficiency; more hydroxyl radicals are generated than with UV and O_3 alone.	Costly, energy-intensive, mass transfer limited, muck production.
O_3	Experience exceptional oxidation performance with minimal time requirements and no residual waste. The residue left by O_3 can be effortlessly decomposed.	Expensive, energy-intensive, and necessitates pre-treatment of compounds.
H_2O_2/UV	Disinfectant, easy procedure.	Turbidity can disrupt the penetration of light, resulting in less efficient generation of •OH.
O_3/H_2O_2/UV	Efficiently breaking down polyphenols and aromatics, without any specific targeting, while considering the entire range of solution species.	Costly, incomplete removal of COD, generation of sludge.
Fenton-based process	The process is economically efficient, environmentally friendly, and has a rapid pace of reaction. It produces potent •OH radicals that can destroy a wide range of contaminants.	Lack of extensive use, little sludge production, and acidic surroundings.
Electro-chemical-based process	Environmentally friendly, capable of automation, functioning as a potent single electron oxidizing agent Demonstrating high effectiveness in generating •OH radicals.	Dependability toward electrode, costly in terms of energy and money.

Source: Prashant Singh, Creative Commons License.

2.7. FUTURE PERSPECTIVES

Learning Objectives:

- Comprehend the future research directions in the field of AOPs.

In the AOPs mentioned earlier, certain aspects have been neglected. For example, when treating underground wastewater, there are various pollutants present, but the focus is primarily on organic pollutants rather than inorganic ones. This can result in a decrease in the overall efficiency of the process, and the water may not be completely free from pollutants. Therefore, it is crucial to focus on enhancing the treatment of inorganic pollutants. In recent times, the utilization of AOPs is rather restricted, primarily confined to laboratory settings or small-scale applications. Efforts must be made to successfully carry out AOPs on a larger scale in order to address the treatment of wastewater and pollutants discharged by textile industries on a significant level (Capodaglio, 2020).

Did You Know?

AOPs have proven to be effective in various applications, including wastewater treatment, air purification, soil remediation, and the advanced oxidation of drinking water.

SUMMARY

- Traditional approaches for reducing industrial pollution were inadequate and unsuitable due to their lack of cost-effectiveness and environmental friendliness. Therefore, it is imperative to find an alternative solution to treat these organic pollutants.
- AOP has emerged as a promising alternative for treating organic pollutants in industrial waste. AOP is a part of the green process that focuses on the degradation of waste released from industries and the treatment of textile wastewater.
- There are various AOPs that have been documented, providing convenience in choosing the most suitable methods for breaking down pollutants. Various AOP methods have been extensively studied to eliminate both organic and inorganic pollutants from wastewater, providing valuable insights into identifying more effective and efficient treatment approaches.
- Based on existing research, it has been observed that photochemical AOPs and Fenton AOP are highly effective in reducing pollutants in textile wastewater. When comparing the effectiveness of UV photolysis or per-sulfate oxidation with UV and radicals, it was discovered that the combination of UV and radicals is more powerful in eliminating pollutants.

REVIEW QUESTIONS

1. What are Advanced Oxidation Processes (AOPs), and what distinguishes them from conventional wastewater treatment methods?
2. Describe the mechanisms involved in ozone-based AOPs for pollutant degradation. How does ozone contribute to the generation of hydroxyl radicals?
3. Explain the principles of AOPs based on UV irradiation. How do photolysis and photodegradation processes contribute to pollutant oxidation?
4. What are catalytic AOPs, and how do catalysts enhance the efficiency of oxidation reactions in wastewater treatment?
5. Explore the various types of catalysts used in catalytic AOPs and their applications in pollutant degradation.

MULTIPLE CHOICE QUESTIONS

1. **Which of the following is a characteristic feature of ozone-based Advanced Oxidation Processes (AOPs)?**
    a. Use of visible light for pollutant degradation
    b. Generation of hydroxyl radicals (•OH) through ozonation reactions
    c. Involvement of electrochemical reactions in pollutant oxidation
    d. Utilization of ultraviolet (UV) irradiation for ozone generation
2. **What is the primary mechanism involved in AOPs based on UV irradiation?**
    a. Electrochemical oxidation
    b. Photolysis and photodegradation
    c. Catalytic degradation
    d. Ozonation reactions
3. **Which type of AOP utilizes physical phenomena such as cavitation and sonolysis for pollutant degradation?**
    a. Ozone-based AOPs
    b. Catalytic AOPs
    c. Physical AOPs
    d. Electrochemical AOPs
4. **What role do catalysts play in catalytic AOPs?**
    a. Generate ozone for pollutant degradation
    b. Enhance the efficiency of oxidation reactions
    c. Facilitate electrochemical oxidation processes
    d. Produce hydroxyl radicals through photolysis
5. **Which electrochemical process is commonly used in electrochemical AOPs for wastewater treatment?**
    a. Ozonation
    b. Electrocoagulation
    c. Photodegradation
    d. Sonolysis

Answers to Multiple Questions

1. (b); 2. (b); 3. (c); 4. (b); 5. (b)

REFERENCES

1. Abbas, I., & Zaheer, S. (2014). Advanced oxidation process for wastewater treatment: A review. *American International Journal of Research in Science, Technology, Engineering, and Mathematics, 7*, 189–191.
2. Ameta, R., Kumar, A., Punjabi, P. B., & Ameta, S. C. (2012). Advanced oxidation processes: Basics and applications. In *Wastewater Treatment: Advanced Processes and Technologies* (pp. 852–855). IWA Publishing.
3. Bahnmüller, S., Loi, C. H., Linge, K. L., Von Gunten, U., & Canonica, S. (2015). Degradation rates of benzotriazoles and benzothiazoles under UV-C irradiation and the advanced oxidation process UV/H_2O_2. *Water Research, 74*, 143–154. https://doi.org/10.1016/j.watres.2014.12.009.
4. Bekbolet, M. (2010). Fundamentals of advanced oxidation processes. In *Water, Wastewater and Soil Treatment by Advanced Oxidation Processes (AOPs)* (Vol. 6, pp. 13–21).
5. Bethi, B., Radhika, G. B., & Sonawane, S. H. (2022). Fundamentals of advanced oxidation processes (AOPs) for wastewater treatment: Challenges and opportunities. In *Novel Approaches Towards Wastewater Treatment and Resource Recovery Technologies* (Vol. 4, pp. 209–220).
6. Capodaglio, A. G. (2020). Critical perspective on advanced treatment processes for water and wastewater: AOPs, ARPs, and AORPs. *Applied Sciences, 10*(13), 4549. https://doi.org/10.3390/app10134549.
7. Chávez, A. M., Gimeno, O., Rey, A., Pliego, G., Oropesa, A. L., Álvarez, P. M., & Beltrán, F. J. (2019). Treatment of highly polluted industrial wastewater by means of sequential aerobic biological oxidation-ozone based AOPs. *Chemical Engineering Journal, 361*, 89–98. https://doi.org/10.1016/j.cej.2018.12.046.
8. Chen, C., Ma, C., Yang, Y., Yang, X., Demeestere, K., Nikiforov, A., & Van Hulle, S. (2023). Degradation of micropollutants in secondary wastewater effluent using nonthermal plasma-based AOPs: The roles of free radicals and molecular oxidants. *Water Research, 235*, 119881. https://doi.org/10.1016/j.watres.2023.119881.
9. Cheng, M., Zeng, G., Huang, D., Lai, C., Xu, P., Zhang, C., & Liu, Y. (2016). Hydroxyl radicals based advanced oxidation processes (AOPs) for remediation of soils contaminated with organic compounds: A review. *Chemical Engineering Journal, 284*, 582–598. https://doi.org/10.1016/j.cej.2015.08.065.
10. Cortez, S., Teixeira, P., Oliveira, R., & Mota, M. (2011). Evaluation of Fenton and ozone-based advanced oxidation processes as mature landfill

leachate pre-treatments. *Journal of Environmental Management, 92*(3), 749–755. https://doi.org/10.1016/j.jenvman.2010.10.006.

11. Guelfi, D. R. V., Gozzi, F., Machulek Jr, A., Sirés, I., Brillas, E., & de Oliveira, S. C. (2018). Degradation of herbicide S-metolachlor by electrochemical AOPs using a boron-doped diamond anode. *Catalysis Today, 313*, 182–188. https://doi.org/10.1016/j.cattod.2018.01.001.
12. Guo, Y., Zhao, E., Wang, J., Zhang, X., Huang, H., Yu, G., & Wang, Y. (2020). Comparison of emerging contaminant abatement by conventional ozonation, catalytic ozonation, O_3/H_2O_2 and electro-peroxone processes. *Journal of Hazardous Materials, 389*, 121829. https://doi.org/10.1016/j.jhazmat.2020.121829.
13. Huang, K. L., Chao, P. J., Kuo, Y. M., Chi, K. Y., Cheng, H. M., Wu, R. Z., & Huang, T. Y. (2021). Response surface methodology-based fabrication of boron-doped diamond electrodes for electrochemical degradation of guaifenesin in aqueous solutions. *Journal of the Taiwan Institute of Chemical Engineers, 123*, 124–133. https://doi.org/10.1016/j.jtice.2021.04.006.
14. Kim, T. H., Kim, S. D., Kim, H. Y., Lim, S. J., Lee, M., & Yu, S. (2012). Degradation and toxicity assessment of sulfamethoxazole and chlortetracycline using electron beam, ozone and UV. *Journal of Hazardous Materials, 227*, 237–242. https://doi.org/10.1016/j.jhazmat.2011.12.038.
15. Kusic, H., Koprivanac, N., & Bozic, A. L. (2006). Minimization of organic pollutant content in aqueous solution by means of AOPs: UV-and ozone-based technologies. *Chemical Engineering Journal, 123*(3), 127–137. https://doi.org/10.1016/j.cej.2006.06.004.
16. Lei, X., Lei, Y., Zhang, X., & Yang, X. (2021). Treating disinfection byproducts with UV or solar irradiation and in UV advanced oxidation processes: A review. *Journal of Hazardous Materials, 408*, 124435. https://doi.org/10.1016/j.jhazmat.2020.124435.
17. Liu, Z., Demeestere, K., & Van Hulle, S. (2021). Comparison and performance assessment of ozone-based AOPs in view of trace organic contaminants abatement in water and wastewater: A review. *Journal of Environmental Chemical Engineering, 9*(4), 105599. https://doi.org/10.1016/j.jece.2021.105599.
18. Ma, D., Yi, H., Lai, C., Liu, X., Huo, X., An, Z., & Yang, L. (2021). Critical review of advanced oxidation processes in organic wastewater treatment. *Chemosphere, 275*, 130104. https://doi.org/10.1016/j.chemosphere.2021.130104.
19. Muruganandham, M., Suri, R. P. S., Jafari, S., Sillanpää, M., Lee, G. J., Wu,

J. J., & Swaminathan, M. (2014). Recent developments in homogeneous advanced oxidation processes for water and wastewater treatment. *International Journal of Photoenergy, 2014*, 8579–8590. https://doi.org/10.1155/2014/857983.

20. Oturan, M. A., & Aaron, J. J. (2014). Advanced oxidation processes in water/wastewater treatment: Principles and applications. A review. *Critical Reviews in Environmental Science and Technology, 44*(23), 2577–2641. https://doi.org/10.1080/10643389.2013.829655.
21. Palaniandy, P., Aziz, H. B. A., & Feroz, S. (2015). Treatment of petroleum wastewater using combination of solar photo-two catalyst TiO_2 and photo-Fenton process. *Journal of Environmental Chemical Engineering, 3*(2), 1117–1124. https://doi.org/10.1016/j.jece.2015.01.009.
22. Pandis, P. K., Kalogirou, C., Kanellou, E., Vaitsis, C., Savvidou, M. G., Sourkouni, G., & Argirusis, C. (2022). Key points of advanced oxidation processes (AOPs) for wastewater, organic pollutants and pharmaceutical waste treatment: A mini review. *ChemEngineering, 6*(1), 8. https://doi.org/10.3390/chemengineering6010008.
23. Pourgholi, M., Jahandizi, R. M., Miranzadeh, M., Beigi, O. H., & Dehghan, S. (2018). Removal of dye and COD from textile wastewater using AOP (UV/O_3, UV/H_2O_2, O_3/H_2O_2 and UV/H_2O_2/O_3). *Journal of Environmental Health and Sustainable Development, 7*(3), 621–629. https://doi.org/10.22034/jehsd.2018.7.3.620.
24. Rana, A. G., Tasbihi, M., Schwarze, M., & Minceva, M. (2021). Efficient advanced oxidation process (AOP) for photocatalytic contaminant degradation using exfoliated metal-free graphitic carbon nitride and visible light-emitting diodes. *Catalysts, 11*(6), 662. https://doi.org/10.3390/catal11060662.
25. Sharma, J., Mishra, I. M., & Kumar, V. (2015). Degradation and mineralization of Bisphenol A (BPA) in aqueous solution using advanced oxidation processes: UV/H_2O_2 and UV/S_2O□^{2}□ oxidation systems. *Journal of Environmental Management, 156*, 266–275. https://doi.org/10.1016/j.jenvman.2015.03.021.
26. Shiying, Y., Ping, W., Xin, Y., Guang, W. E. I., Zhang, W., & Liang, S. H. A. N. (2009). A novel advanced oxidation process to degrade organic pollutants in wastewater: Microwave-activated persulfate oxidation. *Journal of Environmental Sciences, 21*(9), 1175–1180. https://doi.org/10.1016/S1001-0742(08)62471-8.
27. Soon, A. N., & Hameed, B. H. (2011). Heterogeneous catalytic treatment of synthetic dyes in aqueous media using Fenton and photo-assisted

Fenton process. *Desalination, 269*(1–3), 1–16. https://doi.org/10.1016/j.desal.2010.11.001.

28. Tanos, F., Razzouk, A., Lesage, G., Cretin, M., & Bechelany, M. (2024). A comprehensive review on modification of titanium dioxide-based catalysts in advanced oxidation processes for water treatment. *ChemSusChem, 17*(6), 202301139. https://doi.org/10.1002/cssc.202301139.
29. Watts, M. J., Rosenfeldt, E. J., & Linden, K. G. (2007). Comparative OH radical oxidation using UV-Cl_2 and UV-H_2O_2 processes. *Journal of Water Supply: Research and Technology—AQUA, 56*(8), 469–477. https://doi.org/10.2166/aqua.2007.038.
30. Wu, J. J., Wu, C. C., Ma, H. W., & Chang, C. C. (2004). Treatment of landfill leachate by ozone-based advanced oxidation processes. *Chemosphere, 54*(7), 997–1003. https://doi.org/10.1016/j.chemosphere.2003.10.038.
31. Xiao, J., Xie, Y., Rabeah, J., Brüchner, A., & Cao, H. (2020). Visible-light photocatalytic ozonation using graphitic C_3N_4 catalysts: A hydroxyl radical manufacturer for wastewater treatment. *Accounts of Chemical Research, 53*(5), 1024–1033. https://doi.org/10.1021/acs.accounts.9b00620.
32. Yang, K., Lin, H., Jiang, J., Ma, J., & Yang, Z. (2022). Enhanced electrochemical oxidation of tetracycline and atrazine on SnO_2 reactive electrochemical membranes by low-toxic bismuth, cerium doping. *Separation and Purification Technology, 297*, 121453. https://doi.org/10.1016/j.seppur.2022.121453.
33. Zhou, J., Liu, J., Liu, T., Liu, G., Li, J., Chen, D., & Feng, Y. (2023). Electrochemical activation of persulfate by Al-doped blue TiO_2 nanotubes for the multipath degradation of atrazine. *Journal of Hazardous Materials, 445*, 130578. https://doi.org/10.1016/j.jhazmat.2022.130578.
34. Zhuo, Q., Luo, M., Guo, Q., Yu, G., Deng, S., Xu, Z., & Liang, X. (2016). Electrochemical oxidation of environmentally persistent perfluorooctane sulfonate by a novel lead dioxide anode. *Electrochemical Acta, 213*, 358–367. https://doi.org/10.1016/j.electacta.2016.07.102.

CHAPTER 3

OZONE-BASED PROCESSES FOR WASTEWATER TREATMENT

LEARNING OBJECTIVES

At the end of this chapter, readers will be able to:

1. Understand the principles underlying ozone-based AOPs.
2. Comprehend the significance of combining ozone with hydrogen peroxide or ultraviolet light.
3. Gain knowledge about the catalysts use in catalytic ozonation.
4. Learn about ozone by-products formation during ozonation of wastewater.
5. Understand the uses of ozone-based Advanced Oxidation Processes (AOPs) in industrial and municipal wastewater treatment.

INTRODUCTORY EXAMPLE

CleanTech Water Solutions, a leading environmental engineering firm, recently undertook a project to improve wastewater treatment at an industrial facility producing high levels of organic pollutants. The facility, which manufactures chemicals and dyes, was struggling with conventional treatment methods that failed to meet stringent discharge standards. The persistent presence of recalcitrant compounds in the effluent posed a significant environmental risk.

Recognizing the need for an advanced solution, CleanTech Water Solutions proposed the implementation of ozone-based advanced oxidation processes (AOPs). Specifically, they opted for a combination of Ozone/Hydrogen Peroxide (O_3/H_2O_2) and Ozone/UV processes, given their proven efficacy in degrading complex organic molecules.

The O_3/H_2O_2 process was introduced to leverage the synergistic effects of ozone and hydrogen peroxide, generating highly reactive hydroxyl radicals capable of breaking down stubborn pollutants. Additionally, the Ozone/UV process was integrated to further enhance the oxidation potential, utilizing UV radiation to activate ozone and produce more radicals.

The results were impressive. The combined ozone-based AOPs significantly reduced the concentration of organic contaminants in the wastewater. The facility achieved compliance with environmental regulations, ensuring safer discharge into the local water bodies. Furthermore, the treatment process also demonstrated a substantial decrease in the formation of harmful by-products, thanks to the controlled conditions maintained by CleanTech Water Solutions.

UNIT INTRODUCTION

Ozone (O_3) is a highly effective oxidant that has found extensive application in various industries. It is utilized in both industrial and municipal wastewater treatment, disinfection of drinking water, synthesis of chemicals, manufacturing of food and beverages, farming, environmental control, as well as medical and dental fields. Ozone has been utilized for over a century in the United States and globally for the treatment of drinking water. Ozone treatment, also known as ozonation, has been extensively used in the field of wastewater treatment for a multitude of purposes (Haapea et al., 2002). These include effluent sterilization, odor management, color removal, and the oxidation of both inorganic and organic impurities. Ozone is a volatile gas that can be produced through the application of electricity, such as electrical discharge and UV irradiation, to gaseous oxygen molecules. Through a sequence of reactions, ozone in water breaks down into distinct oxidative groups, one of which is the highly potent hydroxyl radical (•OH) (Hussain et al., 2022). Ozonation is known for its powerful oxidizing ability, which is partially due to the production of hydroxyl radicals. Multiple methods can be used to improve the production of hydroxyl radicals in water and wastewater treatment processes using ozone. These techniques encompass the utilization of hydrogen peroxide, metallic catalysts, and UV irradiation. These procedures are widely known as advanced oxidation processes (AOPs) (Rekhate & Srivastava, 2020).

Ozone-based AOPs and ozonation possess the capacity to oxidize a diverse array of organic molecules present in wastewater and water. The substances encompass pharmaceuticals, personal care items, solvents, insecticides, and surfactants. This chapter will examine the basic principles of ozonation and AOPs that utilize ozone in the context of wastewater treatment (Tripathi & Hussain, 2022). It will explore a range of ozone reactions, encompassing both indirect and direct reactions, along with the generation of byproducts and other significant elements to take into account. In addition, it will explore the historical and present studies on ozone-based Advanced Oxidation Processes (AOPs) used in wastewater treatment. This will include a focus on noteworthy initiatives that have specifically utilized ozone for treating wastewater and facilitating water reuse (Kim & Tanaka, 2010) (Figure 3.1).

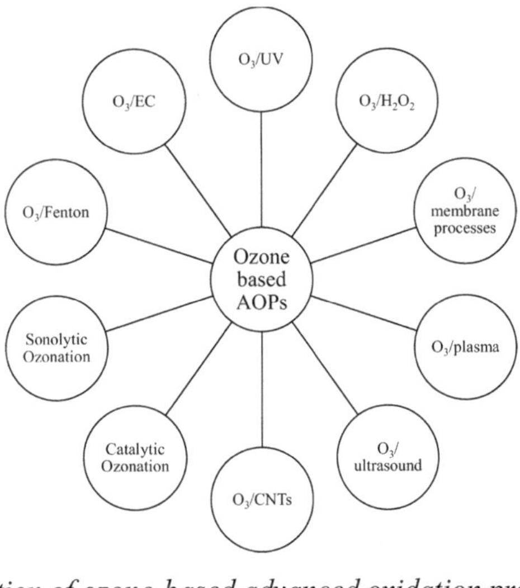

Figure 3.1. *Illustration of ozone-based advanced oxidation processes.*

Source: Sandip Sharma, Creative Commons License.

3.1. OZONE-BASED AOPS

Learning Objectives:

- Gain knowledge about the nature and reactions of ozone.
- Comprehend the mechanisms of different ozone-based AOPs.

Ozone is a gas that reacts easily and has a low solubility in water. Thus, the reactions involving ozone in the aqueous system are quite intricate, encompassing gas-liquid transfer of mass, self-decomposition, and interactions with dissolved and floating/suspended organic and inorganic components. Since ozone has a very high redox potential (E^0) of 2.07 V, it acts as an oxidant when dissolved in water. Ozone reacts with dissolved organic and inorganic substances in a number of ways, such as by indirect hydroxyl radical reactions and direct molecular ozone reactions. It is widely recognized that direct reactions exhibit higher selectivity compared to indirect reactions (Liu et al., 2021).

Beltra´n (2003) categorizes the reactions involving aqueous molecular ozone into three distinct categories:

- Oxidation-reduction reactions;
- Reactions governed by dipolar cycloaddition;
- Reactions governed by electrophilic substitution.

Thanks to its strong redox potential, ozone can easily oxidize a wide range of compounds through oxidation-reduction reactions. These compounds include manganese, iron, nitrite, sulfide, and bromides. Often, these reactions involve the transfer of oxygen from ozone to the reactant. The fundamental half-reactions of ozone within water are (Chávez et al., 2019):

$$O_3 + 2H^+ + 2e^- \rightarrow O_2 + H_2O$$

$$O_3 + H_2O + e^- \rightarrow O_2 + 2OH^- \quad (E^0 = 1.24V)$$

Ozone possesses the capacity to undergo reactions with organic compounds that contain C=C double bonds, such as olefins and aromatics. Ozonide, the cyclic intermediate created by this chemical process, goes through a series of processes, including aberrant ozonolysis. As a result, two smaller molecules—aldehydes, carboxylic acids, or ketones—are produced by the process. Electrophilic substitution involves the reaction of ozone with an organic molecule, including a substitution aromatic compound, in a nucleophilic site (Wu et al., 2004). These compounds often experience ortho- and para-positional hydroxylation as a result of electrophilic substitution processes. When exposed to ozone, hydroxylated phenols degrade and are converted into smaller organic acids, aldehydes, and ketones, through a variety of mechanisms, including aberrant ozonolysis.

Aside from the direct interactions between pollutants and ozone molecules, ozone molecules also undergo decomposition into additional highly reactive oxygen species, often known as free radicals. These free radicals subsequently engage with the toxins present in water. The occurrence of indirect reactions, known as the hydroxyl radical pathway, is primarily driven by the hydroxyl radical, which serves as the main reactive species in these reactions. The production of hydroxyl radicals occurs through a series of radical chain reactions, which encompass initiation, propagation, and termination reactions. There are two primary initiation responses that vary based on the pH levels (Derco et al., 2021):

Acidic to neutral pH:

$$O_3 + OH^- \rightarrow HO_2^{\bullet} + O_2^{\bullet -} \qquad k_{i1} = 70\ M/s$$

Alkaline pH:

$$O_3 + OH^- \rightarrow HO_2^- + O_2 \qquad k_{i2} = 40\ M/s$$

$$O_3 + HO_2^- \rightarrow HO_2^{\bullet} + O_3^{\bullet -} \qquad k_{i3} = 2.2 \times 10^6\ M/s$$

Typically, these initial reactions tend to be the rate-limiting step due to the relatively sluggish reactions of the ozone and hydroxyl ions. Nevertheless, in an alkaline environment, the hydroperoxide ion HO_2^- undergoes a much faster reaction with another ozone molecule, resulting in the formation of a hydroperoxyl radical ($HO_2\bullet$). This radical then breaks down into another free radical, thus continuing the chain reactions:

$$HO_2^{\bullet} \rightarrow O_2^{\bullet -} + H^+ \qquad k_1 = 7.9 \times 10^5 \text{ per s}$$

$$O_2^{\bullet -} + H^+ \rightarrow HO_2^{\bullet} \qquad k_1 = 5 \times 10^{10}\ M/s$$

$$O_3 + O_2^{\bullet -} \rightarrow O_3^{\bullet -} + O_2 \qquad k_2 = 1.6 \times 10^9\ M/s$$

$$O_3^{\bullet -} + H^+ \rightarrow HO_3^{\bullet} \qquad k_3 = 5.2 \times 10^{10}\ M/s$$

$$HO_3^{\bullet} \rightarrow O_3^{\bullet -} + H^+ \qquad k_4 = 3.3 \times 10^2 \text{ per s}$$

$$HO_3^{\bullet} \rightarrow {}^{\bullet}OH + O_2 \qquad k_5 = 1.1 \times 10^5 \text{ per s}$$

$$O_3 + {}^{\bullet}OH \rightarrow HO_4^{\bullet} \qquad k_6 = 2 \times 10^9\ M/s$$

$$HO_4^{\bullet} \rightarrow HO_2^{\bullet} + O_2 \qquad k_7 = 2.8 \times 10^4 \text{ per s}$$

In alkaline environment:

$$HO_2^- + H^+ \rightarrow H_2O_2 \qquad k_8 = 5 \times 10^{10}\ M/s$$

Indeed, hydrogen peroxide plays a crucial role in initiating and promoting the decomposition of ozone, forming the foundation of O_3/H_2O_2 AOPs. According to Andreozzi et al. (1999), the hydroxyl radical has a much greater oxidation potential of 2.8V compared to molecular ozone. This allows it to efficiently and rapidly attack inorganic as well as organic compounds without being selective (Kusic et al., 2006). The rate constants between O_3 and the kOH can reach high values, ranging from 108 to 1,010 M/s. Hence, the elimination of many stubborn organic compounds in water and wastewater is frequently ascribed to the indirect interactions of ozone. Hydroxyl radicals can elicit a range of reactions when they come into contact with organic and inorganic molecules. These reactions encompass electrophilic addition, radical-radical reactions, hydrogen abstraction, and electron transfer reactions. The indirect reactions of ozonation and ozone-based advanced oxidation processes involve not only hydroxyl radicals but also a number of additional extremely reactive oxygen species, including triplet oxygen ($3O_2$), hydroperoxyl radicals, superoxide radical anions, and organic peroxyl radicals.

Various reactions can terminate the indirect ozone reactions. Inhibitors of ozone breakdown have been found as p-chlorobenzoate, tert-butanol, carbonate, bicarbonate, and humic compounds. These chemicals are usually known as hydroxyl radical scavengers. Excessive amounts of hydrogen peroxide can impede the process of ozone breakdown (Sgroi et al., 2021).

Did You Know?

Since the late 19th century, ozone has been utilized to clean water, and it has since become an essential part of advanced oxidation processes (AOPs) used to treat wastewater.

3.1.1. Ozone/Hydrogen Peroxide

The O_3/H_2O_2 advanced oxidation process is widely acknowledged as one of the most thoroughly studied and effectively employed ozone-based AOPs for wastewater and water treatment. In 1982, Staehelin & Hoigne′ undertook a thorough investigation of the O_3/H_2O_2 mechanism. When hydrogen peroxide is present in water, it accelerates the breakdown of ozone, leading to an increased concentration of hydroxyl radicals. The hydroperoxide ion (HO_2^-), a type of ionic hydrogen peroxide, was shown to efficiently initiate free radical reactions when paired with ozone, according to researchers (Tizaoui et al., 2007). Due to the significantly high reaction rate constant (about 106 M/s), the presence of hydrogen peroxide leads to the rapid breakdown of ozone.

Efficient degradation of pollutants can be achieved by utilizing the O_3/H_2O_2 AOP, which facilitates the breakdown of these substances through hydroxyl radical reactions. Research has shown that achieving an optimal ratio of two moles of ozone per mole of hydrogen peroxide is recommended for this purpose (Beltra´n, 2003). In treatment of drinking water, the standard ratios are 0.2–0.5 mg H_2O_2/mg O_3 (Rakness, 2005). In the O_3/H_2O_2 advanced oxidation process, when the target pollutant does not exhibit significant reactivity towards molecular ozone, the majority of contaminant degradation occurs through the indirect chemical processes involving hydroxyl radicals (Beltrán, 2003).

3.1.2. Ozone/UV

Another extensively researched ozone-based AOP is the ozone/UV AOP. The absorption of UV light by dissolved ozone molecules occurs at a wavelength of 260 nm, and it has been observed that the molar absorptivity is 3,292 ± 70 M/cm. When exposed to UV radiation, the dissolved ozone molecules go through photolysis reactions, resulting in the production of hydrogen peroxide (Tichonovas et al., 2017):

$$O_3 + H_2O + h\nu \rightarrow H_2O_2$$

This hydrogen peroxide has the ability to initiate the decomposition of ozone. On the other hand, a molecule of hydrogen peroxide can undergo a different photolysis reaction, resulting in the formation of two hydroxyl radicals.

$$H_2O_2 + h\nu \rightarrow 2{}^{\bullet}OH$$

Furthermore, apart from the reactions involving molecular ozone and hydroxyl radicals, the contaminant under investigation can also undergo photolysis. This occurs when the contaminant absorbs UV light, typically at a wavelength of around 254 nm when low-pressure mercury vapor UV lamps are utilized. It is worth noting that for photolysis to occur, the contaminant must possess a considerable molar absorptivity (Lucas et al., 2010).

3.1.3. Catalytic Ozonation

Several catalytic ozonation processes, both homogeneous and heterogeneous, have been suggested and examined for water treatment. Several transition metal ions, including nickel (II), cobalt (II), copper (II), manganese (II), iron (II), and zinc (II), were discovered to improve the breakdown of contaminants in water. In an acidic medium, Hill (1948) made the initial discovery of ozone decomposition with the assistance of an identical catalyst known as cobalt (II) (Dang et al., 2020).

$$O_3 + Co^{2+} + H_2O \rightarrow {}^{\bullet}OH + CoOH^{2+} + O_2 \quad k = 37 M/min$$

Some have proposed that catalytic ozonation can be regarded as an advanced oxidation process (AOP) since it directly oxidizes cobalt (II) ions with ozone, producing hydroxyl radicals. Furthermore, alongside dissolved metal ions, a range of metal oxides including MnO_2, CuO, TiO_2, and Fe_2O_3, as well as activated carbon and palladium, were discovered to function as heterogenous catalysts for ozone decomposition. Researchers have been actively studying the incorporation of ultraviolet irradiation and H_2O_2 into the catalytic ozonation process, whether it be homogenous or heterogeneous. This particular ozone-based AOP has proven to be effective in treating wastewater, particularly for challenging cases such as high-strength wastewater from industries and landfill leachates with significant amounts of stubborn organic contaminants (Yang et al., 2020).

Practice Problem

Describe the function of catalysts in catalytic oxidation and how they improve the effectiveness of ozone-based Advanced Oxidation Processes (AOPs) in wastewater treatment to degrade organic contaminants.

Solution to Practice Problem

The role of catalysts in Catalytic Ozonation can be explained as follows:
1. Generation of Reactive Oxygen Species (ROS): Catalysts expedite the breakdown of ozone (O3) into very reactive oxygen species, including hydroxyl (•OH) and superoxide radicals (O2•−). These ROS are more potent oxidizing agents than ozone alone and contribute to the degradation of organic pollutants by initiating oxidation reactions.
2. Promotion of Surface Reactions: Catalysts provide active sites on their surface where ozone molecules can adsorb and react with organic pollutants present in the wastewater. This promotes the formation of intermediate products and enhances the efficiency of pollutant degradation by facilitating surface reactions between ozone and organic compounds.
3. Reduction of Activation Energy: Catalysts lower the activation energy barrier for ozone decomposition and pollutant oxidation reactions, leading to increased reaction rates and shorter treatment times. This results in improved overall efficiency and performance of Catalytic Ozonation compared to ozonation without catalysts.
4. Regeneration of Ozone: Certain catalysts possess regenerative properties, allowing them to participate in the ozone decomposition cycle repeatedly without undergoing significant degradation. This ensures the sustained generation of ROS and maintains high oxidation activity throughout the treatment process.

3.2. OZONATION BY-PRODUCTS

Learning Objectives:

- Understand the formation of by-products during ozonation of wastewater.

The presence of small organic molecules during the process of ozonation in the treatment of wastewater and water carries certain implications:

- Production of acids that are organic in nature:
  - Effect on the pH.
- Production of additional biologically degradable organic matter impacts biological stability:
  - Improved efficiency of subsequent biological treatment, particularly in the context of wastewater treatment.
- Production of aldehydes:
  - Aldehydes, for instance, formaldehyde, get regulated.

Furthermore, the process of ozonation results in the formation of two disinfection byproducts: bromate and N-nitrosodimethylamine. The potential carcinogenic effects of both ozonation byproducts have raised concerns among public health experts (Kozari et al., 2020).

3.3. WASTEWATER OZONATION AND OZONE-BASED AOPS

Learning Objectives:

- Understand the main objective of ozonation in wastewater treatment.
- Comprehend the applications of ozone-based AOPs.

Ozone has been widely used in wastewater treatment for a number of years; since the 1970s, both industrial and municipal treatment facilities have included it into their operations. Ozonation plays a crucial role in wastewater treatment by effectively disinfecting and eliminating pathogens. This process is particularly important in municipal wastewater treatment, following biological secondary treatment. Additionally, ozone-based treatment has been thoroughly studied and used for a variety of purposes in diverse locations. It has been tested and utilized for the pre-oxidation process of bio-refractory compounds, enhancement of physicochemical treatment, mitigation of odors, and treatment of sludge (Rodríguez et al., 2008).

3.3.1. Municipal Wastewater Treatment

Effluent disinfection is a common application of ozone in municipal wastewater treatment. The application of ozone-based AOPs in wastewater treatment plants is restricted due to the impact of H_2O_2 and/or UV on the exposure to ozone for disinfection (Jetten et al., 1997). These agents can lead to the breakdown of ozone, which can result in the reduction of the necessary residual concentration and time (Ct). Many researchers have conducted studies on the effectiveness of ozone-based AOPs in eliminating trace organic contaminants found in wastewater and treated effluent. These contaminants include endocrine disruptors, pesticides, medicinal products, and personal care products (Qu et al., 2019).

3.3.2. Industrial Wastewater Treatment

The typical procedures for treating industrial wastewater involve the decomposition of detrimental organic components such as polyaromatic and phenolics hydrocarbons, the elimination of color, and the oxidation of inorganic elements including manganese, iron, sulfides, chromium, and cyanides. Wastewater generally contains substantial quantities of inorganic and organic compounds that exhibit a fast interaction with ozone (Speece, 1983). The compounds mentioned consist of synthetic dyes, phenols, and other additional aromatic substances. In this scenario, the ozone reactions are mostly influenced by mass transfer and are predominantly driven by direct ozone reactions. In addition, there are frequently elevated levels of hydroxyl radical scavengers, such as carbonates and bicarbonate, which can impede the indirect ozone reactions. Multiple investigations have been carried out on advanced oxidation processes (AOPs) that utilize ozone for the purpose of treating different forms of industrial wastewater (Dutta et al., 2021). These investigations involve studying the use of O_3/H_2O_2 and/or O_3/UV AOPs to treat various types of wastewaters, including oil shale wastewater, landfill leachate, pulp mill effluent, textile wastewater, pharmaceutical wastewater, and table olive wastewater. The results suggest that ozone-based Advanced Oxidation Processes (AOP) have a considerable capacity for use in treating industrial wastewater (Westerhoff et al., 2015).

SUMMARY

- Ozone has been used for many years to treat both industrial and municipal wastewater due to its strong disinfection and oxidative properties.
- Ozone-based treatment is exceptionally efficient in treating wastewater containing color derived from both natural (e.g., lignin) aromatic and synthetic (e.g., dyes) compounds. These aromatic compounds possess a high level of resistance to biological oxidation. Ozone possesses the capacity to convert the chemical structures of persistent organic compounds into forms that are more susceptible to decomposition.
- Ozonation is an exceptionally efficient technique for decomposing minute organic compounds present in hospital as well as municipal wastewater. It is becoming more common to employ it in sophisticated water reclamation systems with the goal of producing water drinkable again.
- Despite the slow adoption of ozone-based advanced oxidation processes (AOPs) by full-scale treatment plants, substantial research has been conducted on these technologies for their potential use in wastewater treatment over several years. Given the expenses and technical intricacies associated with ozone treatment, the use of other physical and chemical agents including UV and hydrogen peroxide can substantially augment both the original capital outlay and continuing costs. Many wastewater treatment projects may face difficulties as a result of this.

REVIEW QUESTIONS

1. List down different types of Ozone-based AOPs?
2. What are the primary mechanisms through which ozone acts as an oxidizing agent in wastewater treatment?
3. Describe the principle of Ozone/UV AOP?
4. Describe the role of catalysts in the ozonation process?
5. What are the primary by-products formed during the ozonation of wastewater, and how can they impact water quality??

MULTIPLE CHOICE QUESTIONS

1. **Which of the following is not an ozone-based Advanced Oxidation Process (AOP)?**
    a. Ozone/Hydrogen peroxide
    b. Ozone/UV
    c. Catalytic Ozonation
    d. Electrochemical Oxidation
2. **What is the purpose of combining ozone with hydrogen peroxide in wastewater treatment?**
    a. To enhance disinfection
    b. To increase pH
    c. To improve pollutant degradation
    d. To reduce ozone concentration
3. **Which ozone-based AOP utilizes ultraviolet (UV) light as an additional oxidizing agent?**
    a. Ozone/UV
    b. Ozone/Hydrogen peroxide
    c. Catalytic Ozonation
    d. Ozone alone
4. **What role do catalysts play in Catalytic Ozonation?**
    a. Catalysts neutralize ozone
    b. Catalysts enhance ozone concentration
    c. Catalysts increase pollutant degradation efficiency
    d. Catalysts reduce UV exposure
5. **What is a common by-product formed during ozonation of wastewater?**
    a. Oxygen gas
    b. Chlorine gas
    c. Hydrogen gas
    d. Ozone-depletion compounds

Answers to Multiple Questions

1. (d); 2. (c); 3. (a); 4. (c); 5. (d)

REFERENCES

1. Beltrán, F. J. (2003). Ozone-UV radiation-hydrogen peroxide oxidation technologies. In *Chemical Degradation Methods for Wastes and Pollutants* (Vol. 2, pp. 12–82). CRC Press.
2. Chávez, A. M., Gimeno, O., Rey, A., Pliego, G., Oropesa, A. L., Álvarez, P. M., & Beltrán, F. J. (2019). Treatment of highly polluted industrial wastewater by means of sequential aerobic biological oxidation-ozone based AOPs. *Chemical Engineering Journal, 361*, 89–98. https://doi.org/10.1016/j.cej.2018.12.016.
3. Dang, T. T., Do, V. M., & Trinh, V. T. (2020). Nano-catalysts in ozone-based advanced oxidation processes for wastewater treatment. *Current Pollution Reports, 6*, 217–229. https://doi.org/10.1007/s40726-020-00133-6.
4. Derco, J., Gotvajn, A. Ž., Čižmárová, O., Dudáš, J., Sumegová, L., & Šimovičová, K. (2021). Removal of micropollutants by ozone-based processes. *Processes, 9*(6), 1013. https://doi.org/10.3390/pr9061013.
5. Dutta, D., Arya, S., & Kumar, S. (2021). Industrial wastewater treatment: Current trends, bottlenecks, and best practices. *Chemosphere, 285*, 131245. https://doi.org/10.1016/j.chemosphere.2021.131245.
6. Haapea, P., Korhonen, S., & Tuhkanen, T. (2002). Treatment of industrial landfill leachates by chemical and biological methods: Ozonation, ozonation + hydrogen peroxide, hydrogen peroxide and biological post-treatment for ozonated water. *Ozone: Science & Engineering, 24*(5), 369–378. https://doi.org/10.1080/01919510208908960.
7. Hussain, K., Khan, N. A., Vambol, V., Vambol, S., Yeremenko, S., & Sydorenko, V. (2022). Advancement in ozone-based wastewater treatment technologies: Brief review. *Ecological Questions, 33*(2), 7–19. https://doi.org/10.7206/eq.2353-3508.152.
8. Jetten, M. S., Horn, S. J., & van Loosdrecht, M. C. (1997). Towards a more sustainable municipal wastewater treatment system. *Water Science and Technology, 35*(9), 171–180. https://doi.org/10.2166/wst.1997.0636.
9. Kim, I., & Tanaka, H. (2010). Use of ozone-based processes for the removal of pharmaceuticals detected in a wastewater treatment plant. *Water Environment Research, 82*(4), 294–301. https://doi.org/10.2175/106143009X12583973649150.
10. Kozari, A., Paloglou, A., & Voutsa, D. (2020). Formation potential of emerging disinfection by-products during ozonation and chlorination of sewage effluents. *Science of the Total Environment, 700*, 134449. https://doi.org/10.1016/j.scitotenv.2019.134449.

11. Kusic, H., Koprivanac, N., & Bozic, A. L. (2006). Minimization of organic pollutant content in aqueous solution by means of AOPs: UV-and ozone-based technologies. *Chemical Engineering Journal, 123*(3), 127–137. https://doi.org/10.1016/j.cej.2006.01.033.

12. Liu, Z., Demeestere, K., & Van Hulle, S. (2021). Comparison and performance assessment of ozone-based AOPs in view of trace organic contaminants abatement in water and wastewater: A review. *Journal of Environmental Chemical Engineering, 9*(4), 105599. https://doi.org/10.1016/j.jece.2021.105599.

13. Lucas, M. S., Peres, J. A., & Puma, G. L. (2010). Treatment of winery wastewater by ozone-based advanced oxidation processes (O_3, O_3/UV and O_3/UV/H_2O_2) in a pilot-scale bubble column reactor and process economics. *Separation and Purification Technology, 72*(3), 235–241. https://doi.org/10.1016/j.seppur.2009.12.001.

14. Marrot, B., Barrios-Martinez, A., Moulin, P., & Roche, N. (2004). Industrial wastewater treatment in a membrane bioreactor: A review. *Environmental Progress, 23*(1), 59–68. https://doi.org/10.1002/ep.10178.

15. Qu, J., Wang, H., Wang, K., Yu, G., Ke, B., Yu, H. Q., & Gong, H. (2019). Municipal wastewater treatment in China: Development history and future perspectives. *Frontiers of Environmental Science & Engineering, 13*, 1–7. https://doi.org/10.1007/s11783-019-1141-8.

16. Rekhate, C. V., & Srivastava, J. K. (2020). Recent advances in ozone-based advanced oxidation processes for treatment of wastewater–A review. *Chemical Engineering Journal Advances, 3*, 100031. https://doi.org/10.1016/j.cej.2020.100031.

17. Rice, R. G. (1996). Applications of ozone for industrial wastewater treatment—A review. *Ozone: Science & Engineering, 18*(6), 477–515. https://doi.org/10.1080/01919519608547250.

18. Rodríguez, A., Rosal, R., Perdigón-Melón, J. A., Mezcua, M., Agüera, A., Hernando, M. D., & García-Calvo, E. (2008). Ozone-based technologies in water and wastewater treatment. In *Emerging Contaminants from Industrial and Municipal Waste: Removal Technologies* (Vol. 7, pp. 127–175). CRC Press.

19. Sgroi, M., Anumol, T., Vagliasindi, F. G., Snyder, S. A., & Roccaro, P. (2021). Comparison of the new Cl_2/O_3/UV process with different ozone-and UV-based AOPs for wastewater treatment at pilot scale: Removal of pharmaceuticals and changes in fluorescing organic matter. *Science of the Total Environment, 765*, 142720. https://doi.org/10.1016/j.scitotenv.2020.142720.

20. Speece, R. E. (1983). Anaerobic biotechnology for industrial wastewater treatment. *Environmental Science & Technology, 17*(9), 416–427. https://doi.org/10.1021/es00116a001.
21. Tichonovas, M., Krugly, E., Jankunaite, D., Racys, V., & Martuzevicius, D. (2017). Ozone-UV-catalysis based advanced oxidation process for wastewater treatment. *Environmental Science and Pollution Research, 24*, 17584–17597. https://doi.org/10.1007/s11356-017-0588-2.
22. Tizaoui, C., Bouselmi, L., Mansouri, L., & Ghrabi, A. (2007). Landfill leachate treatment with ozone and ozone/hydrogen peroxide systems. *Journal of Hazardous Materials, 140*(1–2), 316–324. https://doi.org/10.1016/j.jhazmat.2006.08.063.
23. Tripathi, S., & Hussain, T. (2022). Water and wastewater treatment through ozone-based technologies. In *Development in Wastewater Treatment Research and Processes* (Vol. 3, pp. 139–172). Elsevier. https://doi.org/10.1016/B978-0-12-821839-6.00007-4.
24. Westerhoff, P., Lee, S., Yang, Y., Gordon, G. W., Hristovski, K., Halden, R. U., & Herckes, P. (2015). Characterization, recovery opportunities, and valuation of metals in municipal sludges from US wastewater treatment plants nationwide. *Environmental Science & Technology, 49*(16), 9479–9488. https://doi.org/10.1021/acs.est.5b02129.
25. Wu, J. J., Wu, C. C., Ma, H. W., & Chang, C. C. (2004). Treatment of landfill leachate by ozone-based advanced oxidation processes. *Chemosphere, 54*(7), 997–1003. https://doi.org/10.1016/j.chemosphere.2003.11.002.
26. Yang, L., Sheng, M., Li, Y., Xue, W., Li, K., & Cao, G. (2020). A hybrid process of Fe-based catalytic ozonation and biodegradation for the treatment of industrial wastewater reverse osmosis concentrate. *Chemosphere, 238*, 124639. https://doi.org/10.1016/j.chemosphere.2019.124639.

CHAPTER 4

HETEROGENEOUS FENTON-LIKE PROCESSES FOR WASTEWATER TREATMENT

LEARNING OBJECTIVES

At the end of this chapter, readers will be able to:

1. Understand the working mechanism of chemical reactors such as batch reactors.
2. Comprehend the working of different types of continuous reactors such as fixed-bed and fluidized reactor.
3. Understand the principles of heterogeneous Fenton-Like processes and their significance in wastewater treatment.
4. Gain knowledge about the operational parameters that influence the efficiency of heterogeneous Fenton-like processes in continuous reactors.
5. Comprehend the physiochemical properties of heterogeneous catalysts.

INTRODUCTORY EXAMPLE

AquaSolutions Inc., in collaboration with leading research institutions, has developed an advanced wastewater treatment system utilizing heterogeneous Fenton-like processes in continuous reactors. At their state-of-the-art demonstration facility, AquaSolutions Inc. aims to showcase the efficacy of this technology in addressing water pollution challenges and achieving regulatory compliance.

Through the implementation of fixed-bed and fluidized-bed reactors, AquaSolutions Inc. seeks to harness the power of heterogeneous catalysts to facilitate the generation of hydroxyl radicals and degrade organic contaminants present in wastewater streams. By optimizing operational parameters and reactor design, AquaSolutions Inc. aims to achieve high-efficiency pollutant removal while minimizing energy consumption and operational costs.

As AquaSolutions Inc. continues to innovate and refine its wastewater treatment solutions, the company remains committed to promoting sustainable environmental practices and ensuring the availability of clean water for future generations. Through partnerships with industry stakeholders and ongoing research efforts, AquaSolutions Inc. endeavors to make a meaningful impact in the global effort towards water conservation and environmental protection.

UNIT INTRODUCTION

The growing global need for purified water necessitates the creation of economical methods for recycling wastewater. The continuous exploration and advancement of innovative goods, technologies, and manufacturing methods result in the rapid expansion of numerous water-dependent industries. As a result, through several industrial wastewater pathways, new and possibly hazardous compounds are continuously introduced into the environment (Oturan & Aaron, 2014).

Management of pollutants is a very significant topic that has received extensive attention from the scientific community in recent decades. Wastewater discharged by prominent polluting sectors, including chemical (e.g., petrochemical, fertilizer), drugs, textile and dye, food, paper and pulp, and others, frequently contains persistent organic pollutants (POPs) which are resistant to traditional wastewater treatment methods due to their inherent resistance and inability to break down naturally.

In addition, numerous organic compounds pose a potential threat to environment and public health due to their harmful effects on microorganisms, plants, and metabolic activity. These compounds can be toxic, carcinogenic, mutagenic, and disruptors of the endocrine system. Due to rapid industrial development, governments have implemented stricter environmental regulations and health quality standards (Esteves et al., 2019).

Therefore, in various industries, the choice and implementation of a specific treatment process depend on the properties of the wastewater generated, including the type and number of pollutants, and desired reduction level for environmentally safe disposal, while keeping operational and maintenance expenses to a minimum. Various methods for treating wastewater have been extensively researched and applied on an industrial scale. These methods include biological decomposition, the process of precipitation membrane separation, flocculation/coagulation, adsorption, direct oxidation, and combinations of these processes (Xiang et al., 2022).

Most technologies do not include chemical reactions except from biological and chemical processes. Instead, they simply transfer pollutants from one phase to another, resulting in an additional burden on the environment. Nevertheless, in biological processes, they often prove to be inefficient due to the presence of toxic and non-biodegradable molecules in industrial wastewaters (Liu et al., 2019).

Chemical oxidation is often used to break down substances that are resistant to biological degradation, and these techniques can achieve high levels of degradation. However, the process may not be economically feasible due to the requirement for strong oxidants, the equipment needed, and the challenging operating conditions. The exploration of cost-effective and highly efficient methods for treating wastewater has been a captivating subject of conversation (Wang et al., 2016).

Among all the existing treatment methods, advanced oxidation processes (AOPs) can be a valuable technology for breaking down pesticides, aromatics, volatile organics, and solvents, into stable compounds like CO_2, H_2O, and inorganic salts. Even while total mineralization might not be possible, non-biodegradable organic materials can frequently be partially oxidized to yield biodegradable intermediates. This makes it possible to employ later stages of treatment, like biological processes (Hussain et al., 2021).

AOPs have in common the fact that they form and take advantage of the great oxidative capacity of the hydroxyl radical (HO•) which, immediately after the fluorine, is the known species with greater oxidation potential (Table 4.1). In addition, hydroxyl radicals are known for their nonselective and extremely reactive nature, allowing them to effectively target a wide range of organic and inorganic solutes. These radicals exhibit high-rate constants, typically falling within the range of 10^{-6}–10^{-9} L mol^{-1} s^{-1}. The flexibility of AOPs is further enhanced by the availability of various methods for HO• production, including photochemical, chemical, and electrochemical processes. This allows for the selection of the most suitable method to meet specific treatment needs (Rezaei & Vione, 2018).

AOPs can be categorized in a variety of ways according to their reactive phase. One method to classify them is as either heterogeneous, where the metal ions are embedded in the matrix of a solid phase such as carbon nanotubes, activated carbon, clay, or zeolite, or homogeneous, where the catalyst is dissolved in solution (Yamaguchi et al., 2018).

This chapter will explore the extensive research conducted in the past three decades on Fenton/Fenton-like processes, also known as catalytic wet peroxide oxidation (CWPO) or wet hydrogen peroxide catalytic oxidation (WHPCO). The main focus will be on how they can be used in continuous systems to treat simulated chemicals and wastewater from factories (Du et al., 2022).

Table 4.1. *Representation of Oxidation Potential of Commonly Used Oxidants*

Oxidant	Oxidation Potential
Fluorine	3.03
Hydroxyl radical	2.80
Atomic oxygen	2.42
Ozone	2.07
Hydrogen peroxide	1.78
Hydroperoxyl radical	1.70
Potassium permanganate	1.68
Hypochlorous acid	1.49
Chlorine	1.36
Oxygen	1.23
Bromine	1.09

Source: Antonio Gil, Creative Commons License.

Did You Know?

Heterogeneous Fenton-like processes can be tailored to target specific pollutants by selecting appropriate catalyst materials and optimizing operational parameters.

4.1. CHEMICAL REACTORS

Learning Objectives:

- Understand the concept of chemical reactors.
- Comprehend the working procedure of batch reactors.

Chemical reactors, including batch reactors, play a crucial role in chemical engineering processes. They are vital for producing a wide range of chemical compounds, pharmaceuticals, fuels, and for environmental applications like wastewater treatment (Verma et al., 2012).

4.1.1. Batch Reactors

Batch reactors, also known as discontinuous reactors, are frequently utilized for liquid-phase reactions and can use solid catalysts in suspension (also known as slurry reactors). To maintain the ideal temperature inside the reactor, it is crucial to have a recirculation jacket or coil that is linked to a heating or cooling system. Agitation is required in both cases to achieve a uniform solution and maintain consistent temperature throughout the reactor (*see* Figure 4.1) (Cai et al., 2021).

To maintain the ideal temperature in the reactor, it is crucial to have a recirculation jacket or coil linked to a heating/cooling system. Agitation is required in both cases to achieve a uniform solution and maintain consistent temperature throughout the reactor. Therefore, the procedure is conducted in a controlled environment, ensuring that there are no variations in composition and temperature throughout the vessel. However, it should be noted that the composition does change over time. Usually, batch reactors find applications in various industries such as processing and medical product manufacturing. They are commonly used in laboratory settings, especially during the testing phase of a process, or when studying reaction kinetics for modeling and scaling up (de Souza et al., 2020).

Batch reactors are often used in AOP wastewater treatment investigations to evaluate catalysts, fine-tune the process, and analyze the impact of operating parameters. This is particularly evident in the Fenton process. Considering the nature of Fenton's oxidation, it is important to understand that a batch reactor can be seen as a stirring device that operates at moderate temperature and pressure, with no pressurization (Figure 4.2(a)). In an industrial setting, it is necessary to connect it to metering pumps for the addition of acid, base, a hydrogen peroxide solution (35–50% concentration), and a ferrous sulfate catalyst solution (if the process is homogenous). The catalyst is typically maintained in suspension for the heterogeneous Fenton process operation (*see* Figure 4.2(b)) (Herney-Ramírez & Madeira, 2010).

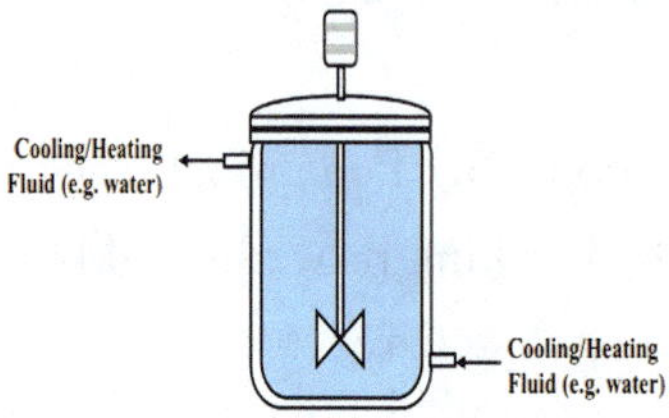

Figure 4.1. *Scheme of a typical batch reactor.*

Source: Antonio Gil, Creative Commons License.

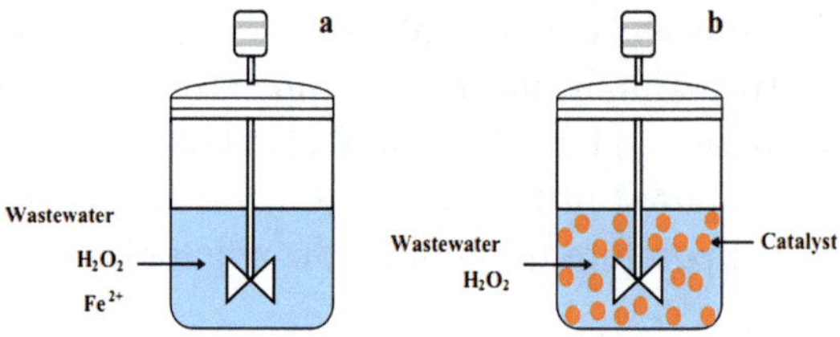

Figure 4.2. *Scheme of wastewater treatment by homogeneous (a); and heterogeneous (b) Fenton process in batch reactors.*

Source: Luis Alejandro, Creative Commons License.

It is important to ensure that the reactor vessel is coated with a material that can withstand the aggressive nature of the Fenton reagent. Corrosion could potentially pose significant operational challenges if not properly addressed. Typically, the reactants are delivered through methods involving the use of diluted sulfuric acid to maintain acidity in the solution, a catalyst in acidic solutions, and a base or acid to adjust the pH. The process of oxidation is started by introducing the oxidant, which may consist of a solution containing hydrogen peroxide. The reaction can be carried out either in a batch reactor-style, where it occurs instantaneously, or in a semi-continuous reactor-style, with a gradual addition over time (Haji et al., 2021).

In an industrial setting, it is necessary to add reactants, especially H_2O_2, gradually while carefully controlling the temperature. Once the solution or wastewater has been processed in the Fenton reactor, it is then discharged into a neutralizing tank in order to adjust its pH levels. Then, it passes through a solid-liquid separation tank and flocculation tank, where the concentration of total dissolved solids (TDS) is frequently regulated. Figure 4.3 shows a Fenton oxidation treatment procedure schematic illustration, which is carried out in the discontinuous mode.

The batch reactor is commonly utilized in laboratory-scale investigations, where both heterogeneous and homogeneous Fenton processes are used to treat model pollutants and various types of wastewaters. In slurry reactors, it is common practice to regularly conduct catalytic effectiveness and stability tests of solid catalysts. This is mainly due to the convenience and ease of operation that these reactors offer. Nevertheless, the implementation of these reactors for industrial applications comes with a set of drawbacks. One of the main challenges is the continuous creation of effluents, which necessitates the use of an equalization tank to hold the effluent before treatment (Chen et al., 2021).

There are certain drawbacks to consider when using the slurry configuration in certain applications. One concern is the high concentration of intermediate products that are formed, which can have a negative impact on the catalyst's longevity. This is because the long contact time between the intermediate products and the catalyst can lead to deactivation mechanisms. In addition, these reactors often incur significant labor costs per unit of production and pose challenges in maintaining large-scale production (Hartmann et al., 2010).

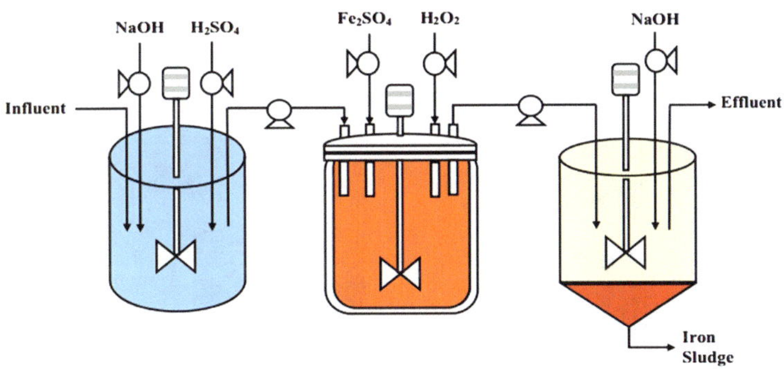

Figure 4.3. *Representative scheme of wastewater treatment by Fenton process.*

Source: Miguel Angel, Creative Commons License.

4.2. CONTINUOUS REACTORS

Learning Objectives:

- Gain knowledge about continuous reactors.
- Comprehend the working mechanism of different types of continuous reactors.

Continuous reactors, often referred to as flow reactors, involve the introduction of reactants to a single location within the vessel, where they then perform a chemical reaction. Afterwards, a steady flow of final product emerges from a different location in the reactor. The continuous stirred-tank reactor (CSTR) and the plug flow reactor are the two main types of continuous reactors. Tubular units frequently use plug flow reactors (Mesquita et al., 2012).

The choice of reactor configuration for executing heterogeneous reactions is determined by several operating characteristics. The fluid phase's velocity is one of them. Typically, two types of reactors are used: packed-bed reactors (PBR) and fluidized-bed reactors (FBR). Using CSTRs with the catalyst suspended, frequently inside a basket, is an additional choice. Here is a brief overview of how heterogeneous Fenton processes typically operate in these reactors (Hodaifa et al., 2013) (Figure 4.4).

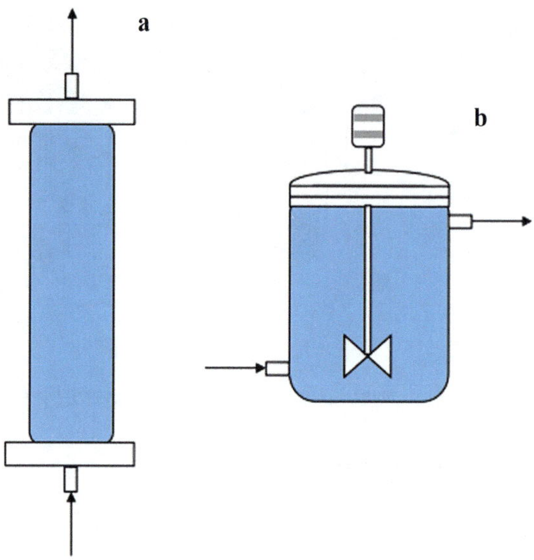

Figure 4.4. *Schematic of a plug flow reactor (a); and a CSTR (b).*

Source: M. Brienza, Creative Commons License.

4.2.1. Fixed-Bed Reactor

Solid catalysts are used in fixed-bed reactors and are arranged in a single bed column (Figure 4.5(a)) or a multi-bed column/multi-tubular reactor (Figure 4.5(b)). The reactor type is referred to as an up-flow fixed-bed reactor, or UFBR, because the reactants are typically introduced into the column in an upward direction.

The oxidant is still consumed, and the solution/effluent is efficiently treated as it passes through the column, despite the less-than-ideal flow pattern where plug flow is not obtained due to variables such axial dispersion effects (Martínez et al., 2007).

The composition (and temperature) in this configuration shows spatial fluctuations both along the axial direction and maybe in the radial direction as well. It is assumed that convective effects alone are responsible for the transfer phenomenon in the idealized and hypothetical plug flow patterns. This indicates that along parallel streamlines, every molecule of the oxidant and effluent travels at the same speed. In research, single-bed reactors have been used to remove colors, oxidize phenolic compounds through heterogeneous Fenton-like reactions, and treat effluent from the pharmaceutical and textile industries (Di Luca et al., 2018).

A smaller number of studies addressing the conjunction of this structure with subsequent biological degradation processes are also mentioned in the literature. As an example, textile effluents are treated with heterogeneous Fenton-like oxidation in a second reactor connected in parallel to the first tubular reactor. Similarly, agrochemical wastewaters are treated with rotating biological contactors following a fixed-bed reactor (Vilardi et al., 2018).

In an industrial setting, this configuration is often used in various chemical reactions that utilize solid catalysts. It finds extensive application in sectors such as the chemical industry (for processes like steam reforming of methane, sulfuric acid production, ammonia synthesis, and methanol synthesis), the petrochemical industry (for the manufacture of vinyl acetate, ethylene oxide, and butadiene), petroleum refining (for polymerization, isomerization, and hydrocracking processes), and fermentation processes (Shao & Chen, 2018).

The fixed-bed reactor's broad industrial applicability can be attributed to its comparatively low running costs and high conversion per unit mass of the catalyst obtained. As far as current understanding allows, fixed-bed reactors have never been used in an industrial setting to treat wastewater using a heterogeneous Fenton-type process. This is most likely connected to the lack of economically viable catalysts (cheap, stable, and non-leaching) for such an application (Botas et al., 2010).

Temperature control is difficult when exothermic events (such the oxidation of highly contaminated wastewater) occur in the reactor due to temperature gradients. Additional disadvantages of this method include mass transfer limits and pressure decreases, which are caused by gas bubbles (likely CO_2 from the oxidation of contaminants inside the column) or the catalyst. Additionally, the deactivation of the catalyst poses an operational challenge when using them in continuous systems, particularly for effluents with complex compositions. In addition, the catalysts utilized in fixed-bed reactors must be molded into spheres, granules, or extrudates to minimize pressure drop throughout the column and ensure optimal mechanical durability (An et al., 2022).

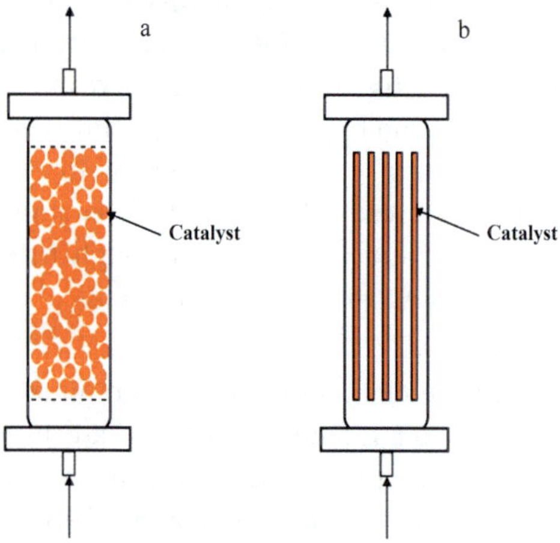

Figure 4.5. *Diagram of a single-bed (a); and multi-bed (b) fixed-bed reactor.*

Source: A. Pintar, Creative Commons License.

4.2.2. Fluidized-Bed Reactor

A liquid phase (or gas) and a solid phase containing tiny particles make up a fluidized-bed reactor (FBR). The structure of this reactor allows the fluid to flow through the catalyst bed, which continuously suspends the solid particles.

This reactor employs a multiphase flow system, specifically either a two-phase or three-phase system, which integrates mass transfer and heat exchange. When a solid is suspended in a fluid stream flowing upwards, it displays characteristics similar to that of a fluid. This improves the system's overall performance and catalytic kinetics. Furthermore, the FBR improves heat and mass transport while avoiding pressure drop-related issues (Tisa et al., 2014).

The catalyst is fluidized through the rapid flow of the fluid within the catalytic bed. The fluid velocity needs to strike a balance between suspending the particles and avoiding the risk of carrying the catalyst out of the reactor. Three different stages may occur based on the fluid velocity and solid distribution in the FBR, as shown in Figure 4.6 (Mohapatra et al., 2021):

1. **Fixed Bed:** The bed demonstrates behavior that is comparable to that of a densely packed bed. The catalyst is closely packed, and the fluid flow rate is at its lowest.
2. **Incipient Fluidization:** The catalyst starts to spread out in the column when the flow rate of the fluid going through the bed of catalyst speeds up. This is where the curve bends, showing the lowest speed needed for fluidization.
3. **Fluidized Bed:** The velocity of fluidization exceeds the minimal threshold, resulting in the expansion of the catalyst particles and their continuous agitation.

Multiple flow regimes can occur, including bubbling fluidization, particle fluidization, turbulent fluidization, slugging fluidization, and pneumatic conveying regimes. These regimes are determined by the fluidization velocity. Fluidized bed reactors (FBRs) have various advantages in comparison to fixed-bed reactors. These benefits encompass superior degradation efficiencies, reduced reaction durations, enhanced recirculation, uniform temperature distribution even for extremely exothermal reactions, as well as improved mixing and contact between liquid and solid substances (Şen & Demirer, 2003).

In the Fenton-like fluidized bed reactor (FBR) operation, the target pollutant crystallizes on the surface of the carrier, reducing the amount of sludge that forms in the homogeneous phase. Fluidized bed reactors (FBRs) find extensive use in the petroleum sector for catalytic cracking, as well as in combustion, gasification, and other catalytic processes. Furthermore, fluidized-bed reactors offer advantages such as ease of building and operation, cost-effectiveness, and the ability to accommodate varying residence durations for both liquid and solid phases. Recently, Fluidized Bed Reactors (FBRs) have been utilized to remove contaminants and treat wastewater (Aghdasinia et al., 2017):

i. Research has been conducted on many methods for the removal of contaminants, including adsorption, biological degradation, and advanced oxidation processes. These advanced oxidation processes include photo-Fenton, electro-Fenton, heterogeneous and homogeneous Fenton-like processes. Additionally, there have been research on combining these processes with adsorption or ultrasound techniques.

ii. Conducting semi-pilot and pilot-scale operations to oxidize dye solutions using Fenton-like processes.

iii. The heterogeneous Fenton-like process is used in industrial applications to treat effluents from industries such as pulp and paper mills, textile dyeing, and petrochemicals (Nejatbakhsh et al., 2023).

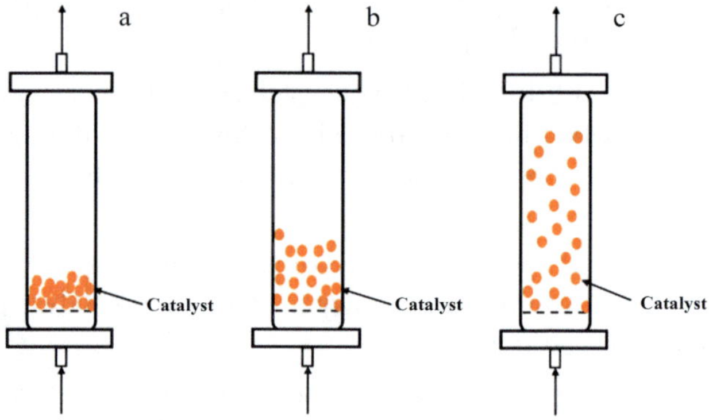

Figure 4.6. *Illustration of fixed-bed (a); incipient fluidization (b); and fluidized-bed (c) stages in an FBR.*

Source: Antonio Gil, Creative Commons License.

Remember

Fixed-bed reactors offer the advantage of having stationary beds of catalyst particles, while fluidized-bed reactors take it a step further by suspending catalyst particles in the reactor through fluidization. This fluidization not only enhances mixing but also improves mass transfer within the system.

4.3. HETEROGENEOUS FENTON/FENTON-LIKE PROCESSES

Learning Objectives:

- Understand the working principle of heterogeneous fenton-like processes.
- Comprehend the operational parameter which influence these processes.
- Learn about the physiochemical properties of heterogeneous catalysts and their impact on catalytic activity and pollutant degradation efficiency.

4.3.1. Fenton's Oxidation

The process of Fenton was discovered by the observation that iron species possess unique oxygen transfer capabilities, which significantly impact the oxidation efficiency of hydrogen peroxide. H. J. Fenton made the initial statement in 1984 during the revision of the oxidation process of tartaric acid using hydrogen peroxide. H. J. Fenton noted that a tiny concentration of ferrous salts (Fe^{2+}) significantly increased H_2O_2 capacity for oxidation. Haber and Weiss proposed in 1934 that the hydroxyl radical is Fenton's highly reactive species (Fu et al., 2022).

The following equation shows how this radical is created when hydrogen peroxide is catalytically broken down in the presence of ferrous ions in an acidic environment:

$$Fe^{2+} + H_2O_2 \rightarrow Fe^{3+} + HO^{\cdot} + OH^{-}$$

An effective method for breaking down organic molecules (RH) in solution is through the use of the Fenton process, which involves the nonselective HO• species. This can be seen in the equation below. There are three ways that oxidation can occur: electron transfer, double bond addition, and hydrogen abstraction. These processes produce fresh oxidized intermediates that gradually break down into H_2O and CO_2 as their molecular weight decreases.

$$HO^{\cdot} + RH \rightarrow H_2O + \text{intermediates}$$
$$HO^{\cdot} + \text{intermediates} \rightarrow H_2O + CO_2$$

While Fenton's oxidation is a complex process that incorporates a mechanism with free radicals and chains, it may be simplified and described by the equations below. The reaction rates for these equations have been well documented in the literature. The reaction linked to the regeneration of the ferrous ion holds great importance as it is accountable for the continuous cycle of the process. The Fenton-like stage, often known as the second equation, is significantly slower and involves the creation of hydroperoxyl radicals (HO_2•). The hydroperoxyl species also react with organic molecules, although their oxidative potential is considerably lower than that of HO•. However, HO_2• has a role in the further restoration of Fe^{2+}, as shown in the calculations below (Jiang et al., 2022).

$$Fe^{3+} + H_2O_2 \leftrightarrow Fe - OOH^{2+} + H^+$$

$$Fe - OOH^{2+} \rightarrow Fe^{2+} + HO_2^{\bullet}$$

$$Fe^{3+} + HO_2^{\bullet} \rightarrow Fe^{2+} + O_2 + H^+$$

Depending on the operational conditions, different competing reactions could happen. Excess H_2O_2 or ferrous ions in solution can scavenge HO• radicals, hence reducing the overall efficiency of the Fenton process. The equations below show these reactions.

$$HO^{\bullet} + H_2O_2 \rightarrow H_2O + HO_2^{\bullet}$$

$$HO^{\bullet} + Fe^{2+} \rightarrow Fe^{3+} + OH^-$$

In addition, there have been reports of radical-radical reactions that can have a negative impact on the process.

$$HO^{\bullet} + HO_2^{\bullet} \rightarrow O_2 + H_2O$$

$$HO^{\bullet} + HO^{\bullet} \rightarrow H_2O_2$$

For oxidation techniques, the Fenton process stands out for its low energy consumption. Unlike other methods, it operates at near-ambient temperature and atmospheric pressure, making it more efficient. One of its advantages is that the activation of hydrogen peroxide does not require any energy input. Additional benefits of the Fenton process are its typically brief reaction times and the utilization of easily manageable chemicals (Kuan et al., 2015).

Numerous studies have highlighted the successful utilization of this process in treating various types of wastewaters, including those from cooking, olive mill, textile, paper and pulp, pharmaceutical, cork-processing industries and cosmetic. Furthermore, it has demonstrated efficacy in the treatment of synthetic wastewaters that comprise certain components such as toluene, phenols, and chlorobenzene, among others.

Despite achieving success in laboratory experiments, the process still needs to be adopted on a larger scale in industrial settings. This is primarily due to (Zhu et al., 2019):

- The limited effectiveness in reducing the levels of certain stubborn organic pollutants (such as carbon tetrachloride, acetic acid, oxalic acid, acetone, trichloroethane, etc.);
- The significant expenses associated with the chemical reagents utilized (such as hydrogen peroxide and pH adjustment chemicals, given the process's narrow optimal pH range of approximately 3.0–3.5);
- The requirement for an additional step to eliminate accumulated iron sludge and partially recover the catalyst, leading to increased overall costs and process complexity;
- The need for high concentrations of a consistent catalyst—iron ions normally range from 50 to 80 parts per million for batch operations—far beyond the 2 parts per million European legal thresholds for the safe release of wastewater into natural water bodies (Bautista et al., 2008).

To address the Fenton process limitation in targeting certain pollutants and minimizing reagent costs, alternative approaches have been suggested. These involve combining Fenton oxidation with established methods such as membrane filtration, biological oxidation, coagulation/flocculation. Combining Fenton's oxidation with other processes is a widely used approach in real industrial wastewater treatment facilities. This is because relying solely on advanced oxidation processes (AOPs) for complete mineralization is not cost-effective and often unnecessary.

Extensive research has been dedicated to exploring the potential of hybrid methods, including electro-Fenton-like, photo-Fenton-like, microwave-Fenton-like, or cavitation-Fenton-like processes, to enhance the Fenton process. In addition, there are other options to consider instead of the traditional Fe^{2+} catalyst. These alternatives include low-valence transition metals like Fe^{3+}, Cu^{2+}/Cu^{+}, and nano-zero-valent iron. These sources are widely recognized for their ability to produce HO• species, which can be utilized in both heterogeneous and homogeneous Fenton-like processes (Babuponnusami & Muthukumar, 2012).

There has been an increasing acknowledgment in recent years of the importance of heterogeneous Fenton-like systems. The utilization of solid catalysts in these systems offers potential solutions to the challenges associated with iron loss and the management of Fe sludge, which are commonly encountered in the homogeneous process. Heterogeneous catalysts can be derived from a variety of sources, including natural iron-bearing materials like pyrite, magnetite, and hematite, as well as artificially synthesized catalysts such as Fe/AC, Fe_2O_3/SBA-15, Fe/ZSM-5, Cu-MCM41, and others. Additionally, industrial by-products like fly ash and sludge can also serve as catalysts. The catalytic activity of these catalysts can be enhanced through various preparation and modification techniques.

Practice Problem

Provide an example of a commonly used heterogeneous catalyst in Fenton-like processes and discuss its physiochemical properties that contribute to enhanced catalytic activity.

Solution to Practice Problem

One commonly used heterogeneous catalyst in Fenton-like processes is iron-based catalysts supported on porous materials such as activated carbon or zeolites. These catalysts possess high surface area and porosity, allowing for efficient adsorption of H_2O_2 and Fe^{2+} ions. Additionally, the presence of iron species on the catalyst surface facilitates the generation of hydroxyl radicals through Fenton-like reactions.
The physiochemical properties of the heterogeneous catalyst, such as surface area, pore size distribution, chemical composition, and morphology, significantly influence its catalytic activity. Higher surface area and pore volume enhance the catalyst's capacity for adsorbing reactants and promoting mass transfer of reactants to active sites. Moreover, the presence of active sites with appropriate redox potentials and coordination environments on the catalyst surface enhances its ability to activate H_2O_2 and Fe^{2+} ions, thereby promoting hydroxyl radical generation and pollutant degradation efficiency.

4.3.2. Operational Parameters of Influence

The chemicals utilized, such as the catalyst and oxidant dosages, are operational parameters that significantly affect Fenton/Fenton-like processes. The pH, temperature, and pollution load—that is, the concentration and properties of the inorganic and organic substances in solution—as well as other reaction parameters are also very important.

Moreover, the quantity of solid catalyst used, which consists of a mixture of low-valence transition metal and support/matrix, as well as the rate at which the feed is supplied, play a critical role in heterogeneous Fenton-like processes conducted in continuous reactors. These factors influence the duration of time the effluent remains in the reactor and the extent of contact it has with the catalyst. Additionally, the characteristics and properties of the catalyst are of utmost importance in this scenario. In order to better demonstrate how these operating factors affect overall process efficiency, a few examples are included in the sections that follow (Papić et al., 2009).

4.3.2.1. Reaction pH

The pH of the reaction medium has a critical role in the efficiency of Fenton/Fenton-like oxidation reactions, particularly in homogeneous systems. This is due to the impact of oxidant speciation variables and catalysts.

For homogeneous systems, it is widely recognized that the Fenton process achieves its best results within a narrow pH range of 3.0–3.5. In these conditions, the optimal concentration of Fe^{2+} active species can be attained while minimizing the undesired breakdown of H_2O_2.

It has been observed that reducing pH levels can have a decreasing effect on the creation of iron (III)-peroxo complexes, which can disrupt the regeneration cycle of Fe^{3+}/Fe^{2+}.

Raising the pH of the reaction leads to the formation of less reactive and insoluble ferric hydroxides (Fe $(OH)_3$) hence decreasing the concentration of ferric ions. In addition, the process of hydrogen peroxide breaking down into molecular oxygen becomes more noticeable in these settings, which hinders the production of oxidizing substances (Ahmed et al., 2021).

$$2H_2O_2 \rightarrow O_2 + 2H_2O$$

One way to address the limitations of narrow pH ranges in homogeneous systems is by using solid catalysts. Heterogeneous Fenton-like catalysts, in particular, have been found to be less affected by pH compared to iron ions in solution, even under similar reaction conditions. Due to the confinement of iron species within the catalyst's structure and pore/interlayer, the formation of iron hydroxide precipitation is minimized, allowing for a sustained generation of HO•.

For example, the altered polyacrylonitrile (PAN) catalyst remains catalytically active within the typical pH range (6.6–8.0) of these effluents. This catalyst was used in a pilot-scale continuous stirred-tank reactor to oxidize pharmaceutical and personal care products (PPCPs) as well as trace endocrine disrupting compounds (EDCs).

According to the scientists, the presence of strong ligands that securely bind the metal ions can lead to catalytic activity over a broader pH range. This is because these ligands inhibit the formation of solid iron particles, which usually happens in uniform systems with a pH above 4. Additionally, a researcher reported that the affordable Fe_2O_3/γ-Al_2O_3 catalyst maintained its catalytic activity within the pH range of 5–9 when utilized in a fixed-bed reactor for the continuous Fenton-like oxidation of a phenolic solution. During those runs, the elimination of COD and phenol was >90% and >99%, respectively (steady-state values), only decreasing when pH was less than 10 (Figure 4.7) (Choquehuanca et al., 2021).

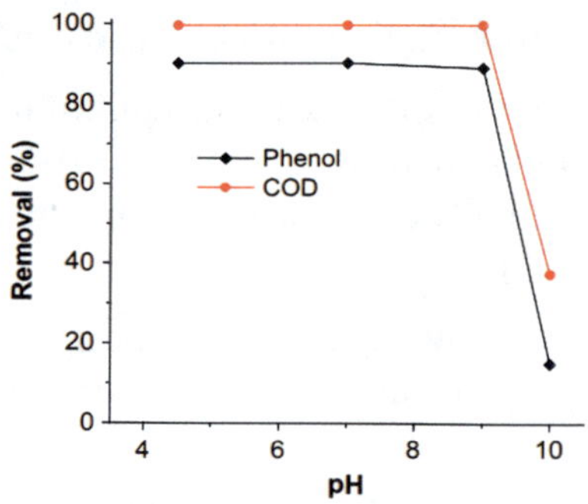

Figure 4.7. *Influence of the initial pH on the degradation of phenolic solutions at steady state.*

Source: Luis Alejandro, Creative Commons License.

4.3.2.2. Temperature

Temperature is critical factor since it impacts oxidation rate. When the temperature of the reaction medium is raised, the oxidation process accelerates. The reason for this phenomenon lies in the strong relationship between kinetic constants (k) and temperature, as described by the Arrhenius law. This relationship applies to various processes such as radical production, iron regeneration, and organics attack.

$$k = k_0 \exp\left(-\frac{E_a}{RT}\right)$$

where; k_0 represents the global rate constant; R stands for the ideal gas constant; E_a denotes activation energy; and temperature is represented by T. The correlation between discontinuous and continuous processes is especially noticeable within the temperature range of 10 to 40/50°C, as documented by multiple researchers. A researcher found that when using a Fe/AC catalyst in a fixed-bed reactor, the decolorization and TOC removal of a dye solution followed the Arrhenius law dependence within the temperature range of 10–50°C (Tolba et al., 2019). The decolorization and TOC degradation reach 74% and 37% respectively at an optimal temperature of 50°C (steady-state values). At a temperature of 70°C, the oxidation of the dye solution did not show any further improvement with an increase in temperature. In fact, there was a minor drop in mineralization efficiency. This occurrence is probably linked to the thermal breakdown of H_2O_2 into oxygen and water. This decomposition is accelerated at higher temperatures, leading to an increase in the production of HO• and subsequently impacting the decay of organic matter (Huang et al., 2010). Furthermore, during operation at T ¼ 70°C, the formation of gas bubbles in the reactor's column would have adversely affected mass transfer from the liquid phase to the catalytic phase, so restricting the process's effectiveness. A similar study was conducted where they also noticed the formation of gas bubbles within the fixed-bed reactor while operating at temperatures ranging from 50 to 70°C. Nevertheless, the prevalence of this occurrence was insufficient to hinder the Chicago Sky Blue (CSB) dye solution's ability to degrade (Figure 4.8).

Similar findings were highlighted regarding the degradation of phenol using a Fe_2O_3/MCM-41 mesoporous zeolite catalyst. The results suggest that the behavior of each catalytic system varies based on the reaction conditions. Hence, it is crucial to optimize every operational variable in order to guarantee a cost-efficient implementation of Fenton-like procedures. In order to enhance degradation rates and improve process efficiency for industrial applications, it may be possible to increase the reaction temperature up to a certain limit. Under certain conditions, such as in the case of effluents from textile industries, it may be possible to eliminate the need for heating the system.

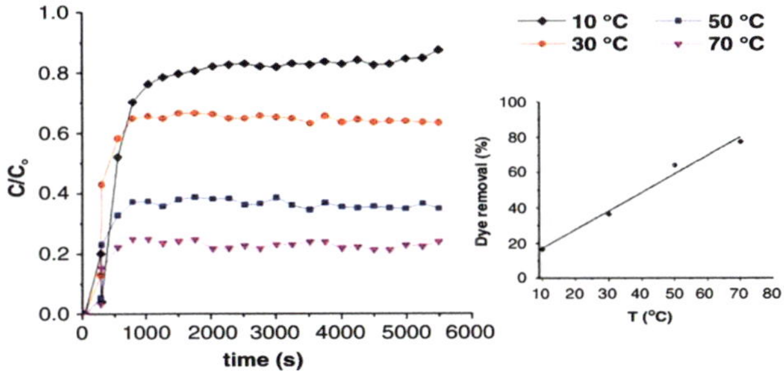

Figure 4.8. *Representation of dye concentration history at the fixed-bed reactor outlet for different reaction temperatures, inset graph shows steady-state dye conversion values.*

Source: Miguel Angel, Creative Commons License.

This is because textile wastewaters often have high temperatures when they are discharged. However, in many cases, the expenses related to energy consumption and the negative impact of leaching on iron loss, particularly at higher temperatures, suggest that lower temperatures (around 30°C) are more appropriate for most industrial wastewater treatment facilities (Shankar et al., 2018).

Similar findings were observed regarding phenol degradation using a Fe_2O_3/MCM-41 mesoporous zeolite catalyst. The results suggest, the behavior of each catalytic system varies based on the reaction conditions. To use Fenton-like processes cost-effectively, each operational variable must be optimized. In order to enhance degradation rates and improve process efficiency for industrial application, it is possible to increase the reaction's temperature up to a certain point.

Under certain conditions, such as with effluents from textile industries, it may be possible to eliminate the need for heating the system. This is because textile wastewaters often have high temperatures when they are discharged. However, in many cases, the expenses related to energy consumption and the negative impact of leaching on iron loss can be significant, especially at higher temperatures. As a result, it is often more practical for most industrial wastewater treatment facilities to operate at lower temperatures, around 30°C (Yao et al., 2018).

4.3.2.3. Pollutant Concentration Fed

Most AOPs can only be used effectively if the target pollutant or wastewater is of a certain type and concentration. At lower concentrations of the target compound(s), high or nearly full mineralization by Fenton/Fenton-like reactions is almost always preferred. Most of the research that has been published in the literature and referenced in this chapter related to the handling of dye solutions and phenols (Lin et al., 2022).

Phenolic compounds are commonly detected in industrial effluent due to their extensive application in the manufacturing of pesticides, lubricating oils, solvents, resins, and dyes. Extended exposure to these chemicals has been linked to human health hazards and microorganism and plant toxicity, even at very low concentrations. Since pigments and dyes are widely employed as coloring ingredients in the food, pharmaceutical, and textile sectors, reports of them are also prevalent.

Because dye-containing effluents, especially those in the azo dye category, include organic chemicals that are difficult to break down using traditional procedures, their discharge is recognized to pose serious environmental risks. It was observed that the Fenton-like oxidation process of Basic Blue 3 dye (BB3) using natural magnetite, increasing the starting concentration of BB3 within the range of –20 mg L^{-1} resulted in a decrease in the efficiency of degradation, as shown in Figure 4.9 (Khataee et al., 2009).

The authors highlight how adjusting the initial dye concentration may have an impact on the function of active areas on the natural magnetite surface. Such a situation can result in the deactivation of available sites for H_2O_2, leading to a reduction in the production of hydroxyl radicals. Research was conducted on semi-continuous catalytic wet phenolic solution oxidation. With certain parameters like temperature, H_2O_2 presence, and catalyst use, they obtained identical results in a specified feed concentration range of 100–2,000 ppm (T ¼ 70°C, in the presence of 0.3 M of H_2O_2 and 6.6 g L^{-1} of an Al-Fe-pillared clay catalyst) (Rene et al., 2012).

After 240 minutes of reaction, the removal of phenol was almost entirely achieved for feed concentrations ranging from 100 to 1,000 ppm. However, it was noted that oxidation rates were slower when the initial phenol concentration was increased within that same range. The degradation efficiency was found to be almost half of what had been previously noted for the same reaction time when the phenol feed dose was raised to 2,000 ppm.

Overall, it has been demonstrated that Fenton-like processes are successful in breaking down stubborn compounds. Nevertheless, there are difficulties that arise when working with substantial quantities of organic matter. This necessitates a notable rise in H_2O_2 consumption and may result in the catalyst's surface becoming obstructed (Waqas et al., 2023).

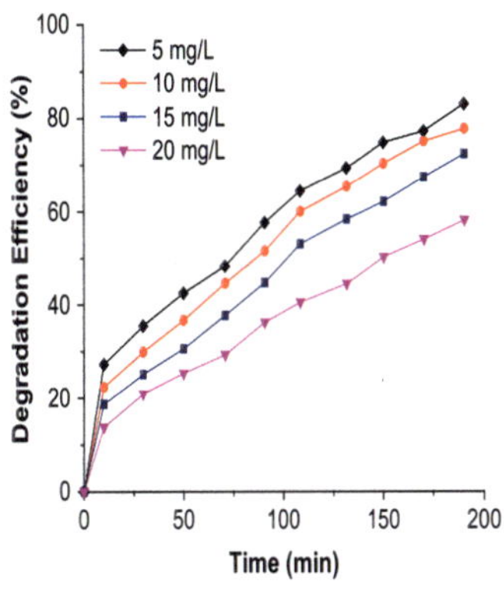

Figure 4.9. *Effect of BB3 initial concentration on the BB3 degradation efficiency in semi-continuous operation.*

Source: M. Sheikhi, Creative Commons License.

4.3.2.4. Catalyst Load

Within Fenton-like processes/heterogeneous Fenton, transition metal ions with low valence, such as ferric or ferrous ions, are involved in the reaction scheme, resulting in hydroxyl radical species production. The metal ions are situated at the surface locations of the solid substance. Just like in homogeneous systems, having a small amount of catalyst in the solution will generally lead to lower process efficiency. This is because there is less catalyst available to react with H_2O_2. Furthermore, in fixed-bed reactors, the residence period (t_R) of the liquid phase is indirectly affected by the catalyst load, as the following equation shows.

$$t_R = \frac{V_L}{Q} = \frac{\epsilon_L V_B}{Q}$$

Where liquid holdup, E_L, is a measure of the relationship between the volume of liquid, V_L, and the volume of the catalyst bed, V_B (sometimes referred to as the height of the catalyst bed). Q, however, represents the overall rate at which feed is flowing. The calculation of the liquid holdup is based on parameters associated with the packed bed, specifically, porosity and external liquid saturation (Ahmed et al., 2016).

It is worth noting that the study conducted by Yan et al. demonstrated that the fluid residence time can be altered by adjusting the catalytic volume/mass. A study was carried out to analyze the influence of the duration of treatment on the process of catalytic wet hydrogen peroxide oxidation of phenol. The height of the catalyst bed was adjusted within a range of 1.1–3.8 cm, which referred to 2.5–9.7 g of Fe/ZSM-5 catalyst, to achieve this (Thomas et al., 2021).

As expected, the longer catalytic bed heights resulted in higher conversions of phenol, TOC, and H_2O_2. The solution's extended residence time or increased active iron species may explain this. As part of their research, Mesquita et al. wanted to enhance the dye conversion of a solution containing an azo dye. To achieve this, they decided to increase the column load by using a Fe/AC catalyst with a higher concentration of Fe (7 wt.%). The column load was raised from 4.7 to 10.25 g.

The increase resulted in around 70% dye conversion under steady state conditions (at a temperature of 30°C, using a feed comprising 0.012 mM of the dye and 2.25 mM H_2O_2). The observed performance boost, in contrast to the approximately 37% improvement achieved with 4.7 g of Fe/AC, can be given to the longer catalyst beds leading to extended contact times, as indicated by the ratio W_{cat}/Q (i.e., the catalyst amount, W_{cat} in the reactor is increased while keeping the feed flow rate, Q, constant) (Ilie et al., 2011).

4.3.3. Catalyst's Physicochemical Properties Influence

An inexpensive, highly catalytically active, and stable material would be ideal for heterogeneous Fenton/Fenton-like reactions. It is claimed that a number of factors directly affect the stability of heterogeneous catalysts and the effectiveness of Fenton's oxidation. These include, among other things, the kind and supply of support, incorporating and synthesizing the catalyst (such as the kind of precursor used and the metal ion chosen as the active phase), and the size of the particles added to the catalytic system.

Rodriguez and his team examined the catalytic efficiency of Fe-based catalysts (containing 4.2–4.9 wt.% of Fe) supported on nine different materials for degrading the acid orange II dye (OII) using a batch reactor in a Fenton-like process. Further research was conducted to determine whether a chemical reaction or adsorption was taking place in the same experimental setup, since the adsorption of organic molecules onto the solid surfaces of the catalysts can improve process efficiency.

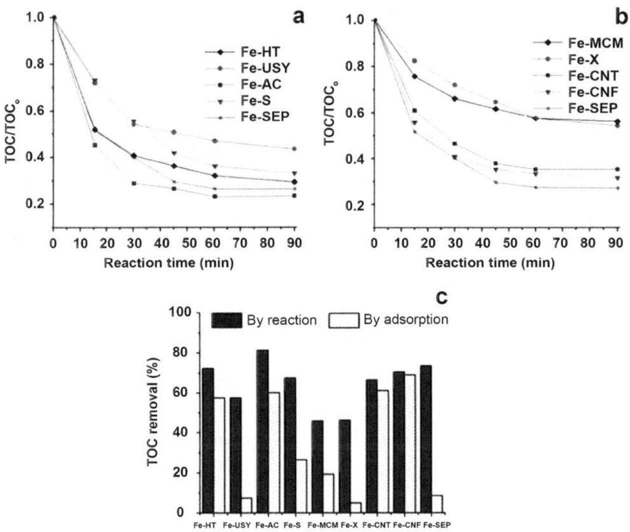

Figure 4.10. *(a), (b) Evolution of TOC along time; and (c) TOC removal after 120 min (by adsorption and heterogeneous Fenton reaction) for different Fe-based material.*

Source: M. Mestanza, Creative Commons License.

The study found that hydrotalcite and carbon (AC, CNT, and CNF) supports were more likely than siliceous and clay mineral supports (such sepiolite) to show an overall TOC decrease of OII solutions by adsorption phenomena (*see* Figure 4.10). Among the nine materials examined, the Fe catalyst based on sepiolite demonstrated the highest viability as a Fenton catalyst, with the lowest percentage of iron leaching. Additionally, it exhibited the highest rates of total organic carbon (TOC) and chemical oxygen demand (COD) conversion, along with the superior overall color (Dihingia & Tiwari, 2022).

4.3.4. Heterogeneous Catalysts Operation Issues: Stability

As mentioned earlier, various materials exhibit varying levels of activity in promoting the Fenton's reaction. Ensuring a strong bond between the active species and the heterogeneous support is crucial for maintaining long-lasting catalytic activity and strength in continuous heterogeneous Fenton-like systems. This is necessary for the successful implementation of these systems on a larger scale. One frequent cause of catalyst deactivation is the migration of metal species from the liquid phase to the solid support. As a homogenous catalyst, a solution with a moderate concentration of the metal species can be used. This suggests that, besides the influence of the heterogeneous catalyst, some of the catalytic activity is also caused by the presence of leached species in the solution. It is important to remember that this could result in secondary metal ions contaminating the treated effluent. The easiest way to find a catalyst's maximum productivity and stability is to use a fixed amount of catalyst and extend the reaction period until the catalyst deactivates entirely or partially. If needed, appropriate reactivation procedures can be developed based on the data collected during the process (Kalz et al., 2017).

Zazo conducted long-term stability studies using a Fe/AC catalyst with a Fe concentration of 4 wt.% in a CSTR. Using a feed flow rate of 5 mL min^{-1}, consisting of solutions containing (500 mg L^{-1} H_2O_2 and 100 mg L^{-1} phenol solutions), the conversion of phenol exceeded 90% within the first 20 hours of the process. Furthermore, around 80% degradation of total organic carbon (TOC) was found within the same period, specifically referring to the values in a steady-state condition. Starting from that moment and continuing until the 170-hour mark, there was a gradual decrease in activity, resulting in removal percentages of approximately 65% for phenol and 25% for TOC. The AC support had a nearly 50% reduction in total iron concentration after 170 hours of reaction.

The research highlights the strong connection between the decrease in activity and the characteristics of the incoming solution or wastewater. When comparing the degradation of an industrial pharmaceutical wastewater to a phenolic solution, it was observed that the loss of activity in the Fe_2O_3/SBA-15 catalyst was less prominent. This was determined by analyzing the parallel

H_2O_2 conversion levels and TOC reduction after 20 hours of reaction. Another factor that can lead to catalyst deactivation is the build-up of carbonaceous deposits, which have been observed to decrease the catalyst's specific surface area, according to research conducted by Zazo et al. (Arends & Sheldon, 2001).

The analysis of the catalyst before and after a 170-hour run in batch operation showed a significant decrease in specific surface area (from 781 to 460 m^2 g^{-1}) and total pore volume (from 0.67 to 0.36 cm^3 g^{-1}). This reduction was observed in both the microporous and all porous structure of the catalyst.

Bruno and Luis implemented reactivation procedures that involved washing the used catalyst with 1 N NaOH. This process successfully recovered most of the SBET and initial pore volume without any loss of Fe. However, the recovered catalyst's catalytic activity showed some irreversible deactivation, as shown by the TOC reduction curves in Figure 4.11.

A common reason for catalyst deactivation is the presence of impurities that can interfere with the active catalytic sites on the surface. This makes the active sites that are responsible for reducing hydrogen peroxide, which is essential for the creation of HO•, inaccessible. During a series of experiments, s gradual decline was noticed in the initial rate of phenol degradation over three consecutive runs. They conducted these experiments using a batch reactor and a Fe-exchanged pillared beidellite catalyst (Campanati et al., 2003).

In addition, the catalyst showed a remarkable recovery of its initial oxidative capability with negligible Fe loss following an intermediate calcination phase. This suggests that the presence of organic species adsorbed on the catalyst's active sites was responsible for the delay in oxidation during short reaction times.

A Fe_2O_3/Al_2O_3 catalyst's stability may be increased by raising the calcination temperature to 900°C and submerging it in an organic acid solution. The interactions between the iron and the support material are improved by these steps.

Various methods have been employed to address the limitations related to the insufficient bonding of metal ions on the catalyst. These methods include altering the surface oxygenated groups through different treatment techniques or using organic ligands to form complexes with iron. These modifications aim to expand the pH range for Fenton oxidation, from acidic to basic. However, the oxidative nature of the process leads to the gradual degradation of the ligands, resulting in the eventual fragmentation of the structure over time.

One significant drawback of Fenton's heterogeneous catalysts is the requirement for pollutant molecules in solution to migrate to the material's surface in order to come into contact with active sites for oxidation. There is a common belief that the rate-limiting processes in heterogeneous catalysts involve surface interactions, as well as the adsorption/desorption of reactants and metal ions. However, the exact mechanism behind these activities remains unknown (Julian et al., 2019).

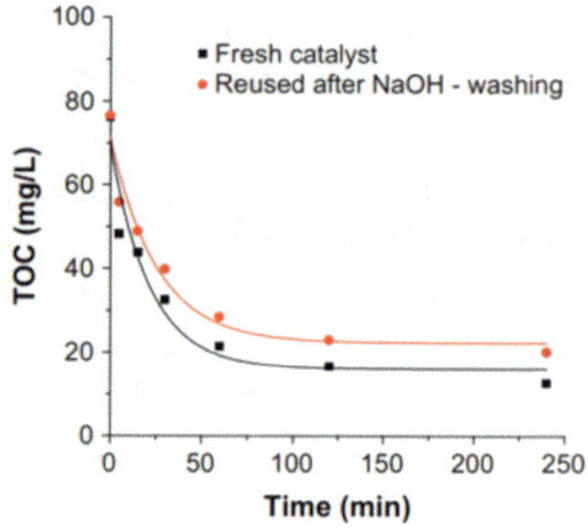

Figure 4.11. *Comparison of fresh vs used catalyst regenerated with 1 N NaOH-solution in batch runs.*

Source: J. A. Zazo, Creative Commons License.

SUMMARY

- There is no doubt that the breakdown of refractory organic compounds in wastewaters produced by industry has increased interest in homogenous Fenton and Fenton-like processes.
- Over past few decades, there has been a significant focus on researching heterogeneous materials that can support the active phase of Fenton's reagent (typically iron species), resulting in high catalytic activity and stability.
- Recently, there has been a growing focus on using diverse catalysts in continuous systems such as fluidized-bed reactors, fixed-bed reactors, and CSTRs. These systems are particularly favored for industrial applications due to their scalability. Each configuration has its own set of pros and cons for applying the heterogeneous Fenton-like process, which will be discussed in detail in this chapter.
- The chapter also highlights the main technological issue related to this process, which is the deactivation of catalysts.
- The concluding part of this chapter shows the capabilities of continuous systems that use heterogeneous Fenton-like reactions. The outcomes of these systems are being compared to those of batch processes that were carried out in similar situations.

REVIEW QUESTIONS

1. Compare and contrast batch reactors and continuous reactors for wastewater treatment, highlighting their advantages and limitations.
2. Describe the operation and configuration of fixed-bed reactors and fluidized-bed reactors used in heterogeneous Fenton-like processes for wastewater treatment.
3. Explain the principle of Fenton's oxidation and how hydroxyl radicals (•OH) are generated in the reaction. Discuss its significance in pollutant degradation.
4. What are the key operational parameters that influence the efficiency of heterogeneous Fenton-like processes in continuous reactors?
5. Identify and discuss the operational issues associated with the use of heterogeneous catalysts in continuous reactors for wastewater treatment.

MULTIPLE CHOICE QUESTIONS

1. **Which type of reactor provides continuous flow operation for wastewater treatment?**
   a. Batch reactor
   b. Fixed-bed reactor
   c. Stirred-tank reactor
   d. Plug-flow reactor
2. **What is the core principle of Fenton's oxidation in wastewater treatment?**
   a. Photocatalytic degradation
   b. Hydrolysis of organic pollutants
   c. Generation of hydroxyl radicals (•OH)
   d. Biological degradation
3. **Which operational parameter significantly influences the efficiency of heterogeneous Fenton-like processes?**
   a. Reactor volume
   b. Catalyst surface area
   c. pH of the solution
   d. Temperature of the reaction
4. **What role do heterogeneous catalysts play in Fenton-like processes?**
   a. They act as oxidizing agents
   b. They facilitate the generation of hydroxyl radicals (•OH)
   c. They enhance mass transfer in the reactor
   d. They neutralize acidic solutions
5. **Which characteristic of catalysts is important for catalyzing Fenton-like reactions?**
   a. High pH
   b. Low surface area
   c. Large pore size
   d. Chemical composition and morphology

Answers to Multiple Questions

1. (b); 2. (c); 3. (c); 4. (b); 5. (d)

REFERENCES

1. Aghdasinia, H., Khataee, A., Sheikhi, M., & Takhtfiroozeh, P. (2017). Pilot plant fluidized□bed reactor for degradation of basic blue 3 in heterogeneous Fenton process in the presence of natural magnetite. *Environmental Progress & Sustainable Energy, 36*(4), 1039–1048. https://doi.org/10.1002/ep.12713.
2. Ahmed, N., Vione, D., Rivoira, L., Carena, L., Castiglioni, M., & Bruzzoniti, M. C. (2021). A review on the degradation of pollutants by Fenton-like systems based on zero-valent iron and persulfate: Effects of reduction potentials, pH, and anions occurring in waste waters. *Molecules, 26*(15), 4584. https://doi.org/10.3390/molecules26154584.
3. Ahmed, Y., Yaakob, Z., & Akhtar, P. (2016). Degradation and mineralization of methylene blue using a heterogeneous photo-Fenton catalyst under visible and solar light irradiation. *Catalysis Science & Technology, 6*(4), 1222–1232. https://doi.org/10.1039/C5CY01681H.
4. An, Z., Hu, Y., Zhang, D., Zhou, H., Zhan, J., & Wu, M. (2022). Preparation of MoS2/SiO2 composites as fixed-bed reactors for Fenton-like advanced oxidation of sulfonamides in water. *Journal of Environmental Chemical Engineering, 10*(3), 107867. https://doi.org/10.1016/j.jece.2022.107867.
5. Arends, I. W. C. E., & Sheldon, R. A. (2001). Activities and stabilities of heterogeneous catalysts in selective liquid phase oxidations: Recent developments. *Applied Catalysis A: General, 212*(1–2), 175–187. https://doi.org/10.1016/S0926-860X(00)00719-8.
6. Babuponnusami, A., & Muthukumar, K. (2012). Advanced oxidation of phenol: A comparison between Fenton, electro-Fenton, sono-electro-Fenton, and photo-electro-Fenton processes. *Chemical Engineering Journal, 183*, 1–9. https://doi.org/10.1016/j.cej.2011.10.058.
7. Bautista, P., Mohedano, A. F., Casas, J. A., Zazo, J. A., & Rodriguez, J. J. (2008). An overview of the application of Fenton oxidation to industrial wastewaters treatment. *Journal of Chemical Technology & Biotechnology: International Research in Process, Environmental & Clean Technology, 83*(10), 1323–1338. https://doi.org/10.1002/jctb.1924.
8. Botas, J. A., Melero, J. A., Martínez, F., & Pariente, M. I. (2010). Assessment of Fe2O3/SiO2 catalysts for the continuous treatment of phenol aqueous solutions in a fixed bed reactor. *Catalysis Today, 149*(3–4), 334–340. https://doi.org/10.1016/j.cattod.2009.10.008.
9. Cai, Q. Q., Lee, B. C. Y., Ong, S. L., & Hu, J. Y. (2021). Fluidized-bed Fenton technologies for recalcitrant industrial wastewater treatment–Recent advances, challenges and perspective. *Water Research, 190*,

116692. https://doi.org/10.1016/j.watres.2020.116692.

10. Campanati, M., Fornasari, G., & Vaccari, A. (2003). Fundamentals in the preparation of heterogeneous catalysts. *Catalysis Today, 77*(4), 299–314. https://doi.org/10.1016/S0920-5861(02)00420-8.

11. Che, Q., Yang, M., Wang, X., Yang, Q., Williams, L. R., Yang, H., & Chen, H. (2019). Influence of physicochemical properties of metal modified ZSM-5 catalyst on benzene, toluene and xylene production from biomass catalytic pyrolysis. *Bioresource Technology, 278*, 248–254. https://doi.org/10.1016/j.biortech.2019.01.070.

12. Chen, Q., Yao, Y., Zhao, Z., Zhou, J., & Chen, Z. (2021). Long term catalytic activity of pyrite in heterogeneous Fenton-like oxidation for the tertiary treatment of dyeing wastewater. *Journal of Environmental Chemical Engineering, 9*(4), 105730. https://doi.org/10.1016/j.jece.2021.105730.

13. Choquehuanca, A., Ruiz-Montoya, J. G., & La Rosa-Toro Gómez, A. (2021). Discoloration of methylene blue at neutral pH by heterogeneous photo-Fenton-like reactions using crystalline and amorphous iron oxides. *Open Chemistry, 19*(1), 1009–1020. https://doi.org/10.1515/chem-2021-0105.

14. de Souza, P. A. L., Camacho, F. G., da Silva, I. R. D. A., Gonçalves, F. F., Benincá, C., & Zanoelo, E. F. (2020). An experimental and modeling study of the chain initiation reaction in heterogeneous Fenton systems with zero valent iron. *Chemical Engineering Journal, 393*, 124665. https://doi.org/10.1016/j.cej.2020.124665.

15. Di Luca, C., Massa, P., Grau, J. M., Marchetti, S. G., Fenoglio, R., & Haure, P. (2018). Highly dispersed Fe3+-Al2O3 for the Fenton-like oxidation of phenol in a continuous up-flow fixed bed reactor. Enhancing catalyst stability through operating conditions. *Applied Catalysis B: Environmental, 237*, 1110–1123. https://doi.org/10.1016/j.apcatb.2018.05.037.

16. Dihingia, H., & Tiwari, D. (2022). Impact and implications of nano catalyst in the Fenton-like processes for remediation of aquatic environment contaminated with micro-pollutants: A critical review. *Journal of Water Process Engineering, 45*, 102500. https://doi.org/10.1016/j.jwpe.2021.102500.

17. Du, X., Wang, S., Ye, F., & Qingrui, Z. (2022). Derivatives of metal-organic frameworks for heterogeneous Fenton-like processes: From preparation to performance and mechanisms in wastewater purification–A mini review. *Environmental Research, 206*, 112414. https://doi.org/10.1016/j.envres.2021.112414.

18. Esteves, B. M., Rodrigues, C. S., & Madeira, L. M. (2019). Wastewater

treatment by heterogeneous Fenton-like processes in continuous reactors. *Applications of Advanced Oxidation Processes (AOPs) in Drinking Water Treatment, 4*(4), 211–255. https://doi.org/10.1201/9780429260705-13.

19. Fu, W., Yi, J., Cheng, M., Liu, Y., Zhang, G., Li, L., & Yang, X. (2022). When bimetallic oxides and their complexes meet Fenton-like process. *Journal of Hazardous Materials, 424*, 127419. https://doi.org/10.1016/j.jhazmat.2021.127419.
20. Haji, S., Sabea, M. R., Aljawder, T., Al-Aradi, M. H., Al-Khateeb, W. M., Elkanzi, E., & Ahmed, S. (2021). Modeling catalyst deactivation in heterogeneous Fenton-like oxidation reactions. *Chemical Engineering Journal, 416*, 128279. https://doi.org/10.1016/j.cej.2021.128279.
21. Hartmann, M., Kullmann, S., & Keller, H. (2010). Wastewater treatment with heterogeneous Fenton-type catalysts based on porous materials. *Journal of Materials Chemistry, 20*(41), 9002–9017. https://doi.org/10.1039/C0JM00743K.
22. Herney-Ramírez, J., & Madeira, L. M. (2010). Use of pillared clay-based catalysts for wastewater treatment through Fenton-like processes. *Pillared clays and related catalysts, 7*(4), 129–165. https://doi.org/10.1080/09373931003697207.
23. Hodaifa, G., Ochando-Pulido, J. M., Rodriguez-Vives, S., & Martinez-Ferez, A. (2013). Optimization of continuous reactor at pilot scale for olive-oil mill wastewater treatment by Fenton-like process. *Chemical Engineering Journal, 220*, 117–124. https://doi.org/10.1016/j.cej.2012.09.078.
24. Huang, Z., Gedalanga, P. B., Asvapathanagul, P., & Olson, B. H. (2010). Influence of physicochemical and operational parameters on Nitrobacter and Nitrospira communities in an aerobic activated sludge bioreactor. *Water Research, 44*(15), 4351–4358. https://doi.org/10.1016/j.watres.2010.05.015.
25. Hussain, S., Aneggi, E., & Goi, D. (2021). Catalytic activity of metals in heterogeneous Fenton-like oxidation of wastewater contaminants: A review. *Environmental Chemistry Letters, 19*(3), 2405–2424. https://doi.org/10.1007/s10311-021-01258-2.
26. Ilie, A., Simoes, M., Baranton, S., Coutanceau, C., & Martemianov, S. (2011). Influence of operational parameters and of catalytic materials on electrical performance of direct glycerol solid alkaline membrane fuel cells. *Journal of Power Sources, 196*(11), 4965–4971. https://doi.org/10.1016/j.jpowsour.2011.02.046.
27. Jiang, Y., Ran, J., Mao, K., Yang, X., Zhong, L., Yang, C., & Zhang, H.

(2022). Recent progress in Fenton/Fenton-like reactions for the removal of antibiotics in aqueous environments. *Ecotoxicology and Environmental Safety, 236*, 113464. https://doi.org/10.1016/j.ecoenv.2022.113464.

28. Julian, I., Pedersen, C. M., Achkasov, K., Hueso, J. L., Hellstern, H. L., Silva, H., & Santamaria, J. (2019). Overcoming stability problems in microwave-assisted heterogeneous catalytic processes affected by catalyst coking. *Catalysts, 9*(10), 867. https://doi.org/10.3390/catal9100867.
29. Kalz, K. F., Kraehnert, R., Dvoyashkin, M., Dittmeyer, R., Gläser, R., Krewer, U., & Grunwaldt, J. D. (2017). Future challenges in heterogeneous catalysis: Understanding catalysts under dynamic reaction conditions. *ChemCatChem, 9*(1), 17–29. https://doi.org/10.1002/cctc.201600843.
30. Khataee, A. R., Vatanpour, V., & Ghadim, A. A. (2009). Decolorization of CI Acid Blue 9 solution by UV/Nano-TiO2, Fenton, Fenton-like, electro-Fenton and electrocoagulation processes: A comparative study. *Journal of Hazardous Materials, 161*(2–3), 1225–1233. https://doi.org/10.1016/j.jhazmat.2008.04.060.
31. Kuan, C. C., Chang, S. Y., & Schroeder, S. L. (2015). Fenton-like oxidation of 4-chlorophenol: Homogeneous or heterogeneous? *Industrial & Engineering Chemistry Research, 54*(33), 8122–8129. https://doi.org/10.1021/acs.iecr.5b01652.
32. Lin, R., Li, Y., Yong, T., Cao, W., Wu, J., & Shen, Y. (2022). Synergistic effects of oxidation, coagulation and adsorption in the integrated Fenton-based process for wastewater treatment: A review. *Journal of Environmental Management, 306*, 114460. https://doi.org/10.1016/j.jenvman.2021.114460.
33. Liu, Y., Cheng, M., Liu, Z., Zeng, G., Zhong, H., Chen, M., & Song, B. (2019). Heterogeneous Fenton-like catalyst for treatment of rhamnolipid-solubilized hexadecane wastewater. *Chemosphere, 236*, 124387. https://doi.org/10.1016/j.chemosphere.2019.124387.
34. Martínez, F., Melero, J. A., Botas, J. Á., Pariente, M. I., & Molina, R. (2007). Treatment of phenolic effluents by catalytic wet hydrogen peroxide oxidation over Fe2O3/SBA-15 extruded catalyst in a fixed-bed reactor. *Industrial & Engineering Chemistry Research, 46*(13), 4396–4405. https://doi.org/10.1021/ie0700916.
35. Mesquita, I., Matos, L. C., Duarte, F., Maldonado-Hódar, F. J., Mendes, A., & Madeira, L. M. (2012). Treatment of azo dye-containing wastewater by a Fenton-like process in a continuous packed-bed reactor filled with activated carbon. *Journal of Hazardous Materials, 237*, 30–37. https://doi.org/10.1016/j.jhazmat.2012.06.057.

36. Mohapatra, T., Kumar, V., Sharma, M., & Ghosh, P. (2021). Hybrid Fenton oxidation processes with packed bed or fluidized bed reactor for the treatment of organic pollutants in wastewater: A review. *Environmental Engineering Science, 38*(6), 443–457. https://doi.org/10.1089/ees.2020.0188.

37. Nejatbakhsh, S., Soodmand, A. M., Azimi, B., Farshchi, M. E., Aghdasinia, H., & Kazemian, H. (2023). Semi-pilot scale fluidized-bed reactor applied for the Azo dye removal from seawater by granular heterogeneous Fenton catalysts. *Chemical Engineering Research and Design, 195*, 1–13. https://doi.org/10.1016/j.cherd.2022.11.002.

38. Neyens, E., & Baeyens, J. (2003). A review of classic Fenton's peroxidation as an advanced oxidation technique. *Journal of Hazardous Materials, 98*(1–3), 33–50. https://doi.org/10.1016/S0304-3894(02)00282-0.

39. Oturan, M. A., & Aaron, J. J. (2014). Advanced oxidation processes in water/wastewater treatment: Principles and applications. A review. *Critical Reviews in Environmental Science and Technology, 44*(23), 2577–2641. https://doi.org/10.1080/10643389.2013.829040.

40. Papić, S., Vujević, D., Koprivanac, N., & Šinko, D. (2009). Decolourization and mineralization of commercial reactive dyes by using homogeneous and heterogeneous Fenton and UV/Fenton processes. *Journal of Hazardous Materials, 164*(2–3), 1137–1145. https://doi.org/10.1016/j.jhazmat.2008.08.002.

41. Reddy, B. M., Chowdhury, B., & Smirniotis, P. G. (2001). An XPS study of La_2O_3 and In_2O_3 influence on the physicochemical properties of MoO_3/TiO_2 catalysts. *Applied Catalysis A: General, 219*(1–2), 53–60. https://doi.org/10.1016/S0926-860X(01)00797-2.

42. Rene, E. R., Mohammad, B. T., Veiga, M. C., & Kennes, C. (2012). Biodegradation of BTEX in a fungal biofilter: Influence of operational parameters, effect of shock-loads and substrate stratification. *Bioresource Technology, 116*, 204–213. https://doi.org/10.1016/j.biortech.2012.03.031.

43. Rezaei, F., & Vione, D. (2018). Effect of pH on zero valent iron performance in heterogeneous Fenton and Fenton-like processes: A review. *Molecules, 23*(12), 3127. https://doi.org/10.3390/molecules23123127.

44. Şen, S., & Demirer, G. N. (2003). Anaerobic treatment of real textile wastewater with a fluidized bed reactor. *Water Research, 37*(8), 1868–1878. https://doi.org/10.1016/S0043-1354(02)00547-1.

45. Shankar, V. K., Kunar, B. M., & Murthy, C. S. (2018). Experimental investigation and statistical analysis of operational parameters on temperature rise in rock drilling. *International Journal of Heat and*

Technology, 36(4), 1174–1180. https://doi.org/10.18280/ijht.360408.

46. Shao, Y., & Chen, H. (2018). Heterogeneous Fenton oxidation of phenol in fixed-bed reactor using Fe nanoparticles embedded within ordered mesoporous carbons. *Chemical Engineering Research and Design, 132*, 57–68. https://doi.org/10.1016/j.cherd.2018.01.014.
47. Thomas, N., Dionysiou, D. D., & Pillai, S. C. (2021). Heterogeneous Fenton catalysts: A review of recent advances. *Journal of Hazardous Materials, 404*, 124082. https://doi.org/10.1016/j.jhazmat.2020.124082.
48. Tisa, F., Raman, A. A. A., & Daud, W. M. A. W. (2014). Applicability of fluidized bed reactor in recalcitrant compound degradation through advanced oxidation processes: A review. *Journal of Environmental Management, 146*, 260–275. https://doi.org/10.1016/j.jenvman.2014.07.023.
49. Tolba, A., Alalm, M. G., Elsamadony, M., Mostafa, A., Afify, H., & Dionysiou, D. D. (2019). Modeling and optimization of heterogeneous Fenton-like and photo-Fenton processes using reusable Fe_3O_4-MWCNTs. *Process Safety and Environmental Protection, 128*, 273–283. https://doi.org/10.1016/j.psep.2019.05.019.
50. Verma, A. K., Dash, R. R., & Bhunia, P. (2012). A review on chemical coagulation/flocculation technologies for removal of colour from textile wastewaters. *Journal of Environmental Management, 93*(1), 154–168. https://doi.org/10.1016/j.jenvman.2011.09.012.
51. Vilardi, G., Ochando-Pulido, J. M., Stoller, M., Verdone, N., & Di Palma, L. (2018). Fenton oxidation and chromium recovery from tannery wastewater by means of iron-based coated biomass as heterogeneous catalyst in fixed-bed columns. *Chemical Engineering Journal, 351*, 1–11. https://doi.org/10.1016/j.cej.2018.05.057.
52. Wang, N., Zheng, T., Zhang, G., & Wang, P. (2016). A review on Fenton-like processes for organic wastewater treatment. *Journal of Environmental Chemical Engineering, 4*(1), 762–787. https://doi.org/10.1016/j.jece.2015.11.017.
53. Waqas, S., Harun, N. Y., Sambudi, N. S., Abioye, K. J., Zeeshan, M. H., Ali, A., & Alsaadi, A. S. (2023). Effect of operating parameters on the performance of integrated fixed-film activated sludge for wastewater treatment. *Membranes, 13*(8), 704. https://doi.org/10.3390/membranes13080704.
54. Xiang, Y., Liu, H., Zhu, E., Yang, K., Yuan, D., Jiao, T., & Tang, S. (2022). Application of inorganic materials as heterogeneous cocatalyst in Fenton/Fenton-like processes for wastewater treatment. *Separation and Purification Technology, 295*, 121293. https://doi.org/10.1016/j.

seppur.2022.121293.

55. Yamaguchi, R., Kurosu, S., Suzuki, M., & Kawase, Y. (2018). Hydroxyl radical generation by zero-valent iron/Cu (ZVI/Cu) bimetallic catalyst in wastewater treatment: Heterogeneous Fenton/Fenton-like reactions by Fenton reagents formed in-situ under oxic conditions. *Chemical Engineering Journal, 334*, 1537–1549. https://doi.org/10.1016/j.cej.2017.10.045.

56. Yao, D., Zhang, Y., Williams, P. T., Yang, H., & Chen, H. (2018). Co-production of hydrogen and carbon nanotubes from real-world waste plastics: Influence of catalyst composition and operational parameters. *Applied Catalysis B: Environmental, 221*, 584–597. https://doi.org/10.1016/j.apcatb.2017.10.020.

57. Zhu, Y., Zhu, R., Xi, Y., Zhu, J., Zhu, G., & He, H. (2019). Strategies for enhancing the heterogeneous Fenton catalytic reactivity: A review. *Applied Catalysis B: Environmental, 255*, 117739. https://doi.org/10.1016/j.apcatb.2019.117739.

CHAPTER 5

PHOTO CATALYSIS AND WASTEWATER TREATMENT

LEARNING OBJECTIVES

At the end of this chapter, readers will be able to:

1. Understand the principles of photo catalysis.
2. Identify the different types of semiconductor photo catalysts used in photo catalysis.
3. Gain a thorough understanding of the techniques employed to enhance the photo catalytic efficiency of semiconductor photo catalysts.
4. Understand the various applications of photo catalysis used in wastewater treatment.

INTRODUCTORY EXAMPLE

EcoClean Solutions Inc. recently partnered with a pharmaceutical manufacturing plant facing significant challenges in treating its wastewater containing high levels of organic pollutants. Traditional treatment methods were insufficient to meet regulatory standards, leading to concerns about environmental impact and compliance.

To address these challenges, EcoClean Solutions proposed the implementation of photocatalysis using titanium dioxide (TiO_2) photocatalysts to degrade the complex organic compounds present in the pharmaceutical wastewater. By harnessing the power of solar energy and the photocatalytic properties of TiO_2, EcoClean Solutions aimed to provide a sustainable and effective solution for wastewater treatment.

Through a pilot-scale demonstration at the pharmaceutical plant, EcoClean Solutions showcased the efficacy of photocatalysis in degrading organic pollutants, improving water quality, and ensuring regulatory compliance. The successful implementation of photocatalysis not only addressed the wastewater treatment challenges faced by the pharmaceutical plant but also demonstrated the potential of advanced water treatment technologies in achieving environmental sustainability.

As EcoClean Solutions continues to innovate and optimize photocatalysis for wastewater treatment applications, the company remains committed to providing cost-effective and environmentally friendly solutions to industries worldwide, contributing to the preservation of water resources and environmental health.

UNIT INTRODUCTION

The rate of industrialization, technological advancement, and the use of nonrenewable resources has been rapidly increasing in recent decades due to the growing demand for materials used in textiles, dyes, fertilizers, household products, plastics, and other related industries. Therefore, environmental degradation and energy issues are currently at a critical level. Industrial wastes are more harmful and non-biodegradable than municipal wastes due to their composition of fats, grease, petroleum products, phenols, heavy metals, and ammonia, among other substances (Mandade, 2021). Agricultural and pharmaceutical waste discharges contain pesticides and additional chemicals that contribute to the development of chronic diseases and have detrimental effects on the human endocrine system. Consequently, the water becomes unsuitable for consumption and other purposes. There is a pressing requirement to develop innovative technologies that are environmentally beneficial and result in the degradation or complete removal of environmental pollutants, so serving as a viable clean alternative (Ahmed et al., 2010).

Advanced oxidation processes (AOPs) represent an eco-friendly approach for the elimination of a wide spectrum of contaminants, encompassing atmospheric pollutants and diverse forms of water pollutants including, hydrocarbons of petroleum origin, aromatics, dyes, chemical pesticides, insecticides, volatile organic compounds, chlorinated hydrocarbons, and assorted organic compounds. AOPs depend on the generation of reactive oxygen species, specifically hydroxyl radicals, which are distinguished by the presence of an unpaired electron (Petit et al., 2016). These species promptly interact with a broad array of chemical substances that are typically difficult to degrade. AOPs excel over conventional techniques owing to their ability to yield thermodynamically stable oxidation byproducts, including water, carbon dioxide, and biodegradable organic compounds. Within the realm of AOPs lies photo catalysis, which harnesses sunlight through a photo catalyst. These photo catalysts have proven effective in addressing environmental contamination and energy concerns by harnessing diverse segments of the solar spectrum (Meng & Zhang, 2016).

5.1. CONCEPT OF PHOTO CATALYSIS

Learning Objectives:

- Understand the terms photo catalyst and photo catalysis.
- Identify the two types of photo catalytic reactions.
- Comprehend the four ways in which substrate and semiconductor interact with each other.
- Understand the properties of binary, ternary and quaternary oxides.

Photo catalysts are substances that can modify the rate of a chemical reaction when they come into contact with light. This phenomenon is commonly referred to as photocatalysis. Using a semiconductor and light, photocatalysis is a technique that starts chemical reactions. A substrate that both absorbs light, and speeds up chemical reactions is called a photocatalyst.

It is important to note that all photocatalysts can be classified as semiconductors. Photo catalysis is the process by which a semiconducting material generates an electron-hole pair when exposed to light (Guo et al., 2019).

The photo catalytic processes can be classified into two categories based on the physical condition of the reactants:

1. **Homogeneous Photo-Catalysis:** The photo catalytic reactions where both the reactant and semiconductor exist in the same phase, such as liquid, gas, or solid.
2. **Heterogeneous Photo-Catalysis:** The photo catalytic reactions that occur when the reactant and semiconductor are not in the same state.

The energy difference between the valence band (HOMO) and the conduction band (LUMO) is known as the band gap (Eg). Materials are categorized into three basic groups (Figure 5.1) based on their band gap (Ravelli et al., 2009):

- **Conductor:** With Eg equals to 1.0 eV.
- **Semiconductor:** With Eg equals to 1.5–3.0 eV.
- **Insulator:** With Eg less than 5.0 eV.

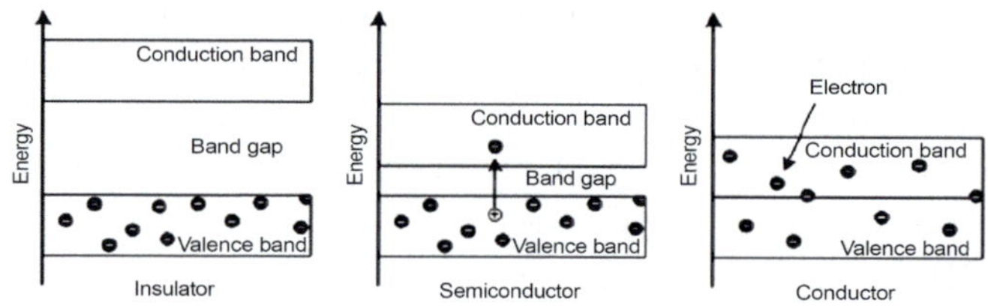

Figure 5.1. *Illustration of different types of materials.*

Source: Suresh Ameta, Creative Commons License.

Semiconductors possess the capacity to conduct electricity at ambient room temperature upon exposure to light, rendering them very efficient as photo catalysts. When a photo catalyst is exposed to light of a suitable wavelength, the electrons in the valence band absorb the energy from the photons and become stimulated, transitioning to the conduction band. This process leads to the formation of a photo-excited state, which produces a pair of an electron (e^-) and a hole (h^+) (Ohtani, 2011). The excited electron is used to decrease the concentration of an acceptor, while a hole is used to increase the concentration of donors by oxidation. The importance of photo catalysis arises from the photo catalyst's capacity to simultaneously generate an oxidation and a reduction environment (Wang et al., 2006).

Did You Know?

Some photocatalysts are naturally occurring minerals, while others are synthesized in laboratories to optimize their photocatalytic activity for specific applications.

The final result of the excited electron and hole depends on the precise arrangement of the valence and conduction bands of the semiconductor, along with the redox potentials of the substrate. The relationship between the semiconductor and the substrate material can be observed in four distinct modes, depending on the positioning of the conduction and valence bands, as well as the redox potentials. These four distinct combinations are illustrated in Figure 5.2:

- When the substrate's redox potential is lower than the conduction band of the semiconductor, substrate reduction takes place.
- If the redox potential of the substrate exceeds the valence band of the semiconductor, the substrate will undergo oxidation.
- When the redox potential of the substrate is higher than the conduction band but lower than the valence band of the semiconductor, neither oxidation nor reduction takes place.
- When the redox potential of the substrate falls within a specific range, a combination of oxidation and reduction processes occur within it (Li et al., 2020).

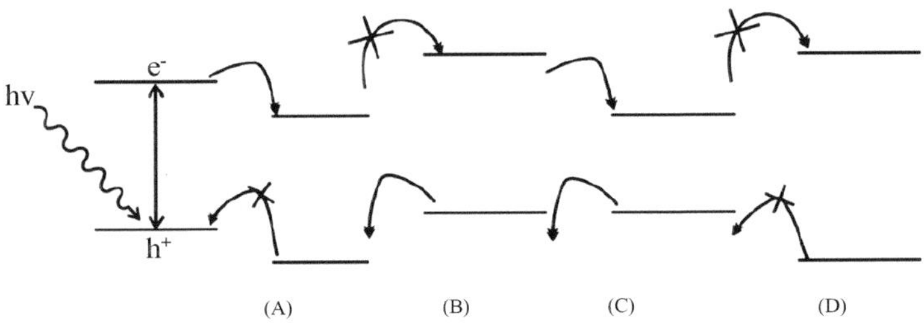

Figure 5.2. *Representation of different possibilities of reactions. (A) Reduction; (B) oxidation; (C) redox reaction; and (D) no reaction.*

Source: Rakshit Ameta, Creative Commons License.

Photocatalysts have a wide range of applications including antifouling, antifogging, energy saving and storage, odor removal, sanitation, self-cleaning, purification of air, and wastewater treatment. Semiconductors serve as stimulants for photoredox procedures owing to their electronic configuration. Certain semiconductors have the ability to catalyze the full mineralization of certain organic contaminants, such as halo hydrocarbons, aromatics, pesticides, insecticides, pigments, and surfactants (*see* Figure 5.3) (Camacho-Muñoz et al., 2020).

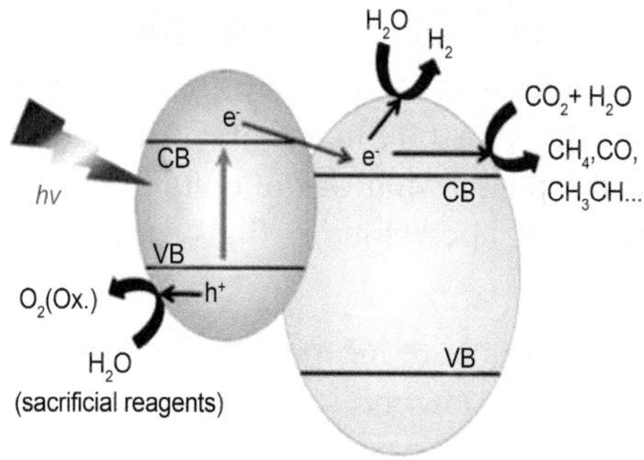

Figure 5.3. *Degradation of organic pollutants.*

Source: Surbhi Benjamin, Creative Commons License.

Metal oxides provide a diverse array of uses in addressing environmental issues and in the field of electronics, owing to their ability to generate charge carriers upon exposure to light.

Metal oxides possess characteristics such as:

- A required electronic configuration;
- Properties related to the absorption of light;
- Charge transport.

The utilization of semiconductors in photocatalysis has garnered significant interest due to its ability to address the issue of rapid recombination of charges (Ohtani, 2010).

5.1.1. Binary Oxides

Photocatalysts are classified into three distinct categories namely Zinc oxide, Titanium dioxide, and oxides of metal such as vanadium, molybdenum, indium, cerium, and tungsten (Chen et al., 2020).

In recent decades, there has been a significant body of research focusing on the application of various binary metal oxides photocatalysts, including ZnO, Fe_2O_3, TiO_2, V_2O_5, CdO, Al_2O_3, and CdS, for the degradation of organic chromogenic contaminants such as azo dyes, acid orange G, alizarin S, methylene blue, Congo red, methyl red, and orange G. The conversion of organic pollutants into NH_4^+, NO_3^-, CO_2, and SO_4^{2-} occurs through the generation of reactive species such as HOO*, O_2^-, or OH* via a series of reduction reactions (Dlamini et al., 2011). The ESR spin-trapping technique, as established by Liu et al. in the year 2000, can be used to identify this mechanism. TiO_2 was used to completely oxidize, detoxify, and fully mineralize many dyes, including methyl red, alizarin S, Congo red, crocein orange G, and methylene blue, in water solutions when exposed to UV light. ZnO and TiO_2 nanoparticles are produced using a nonhydrolytic technique and their efficacy is assessed in the degradation of methylene blue. The reported increase in efficiency was attributed to the existence of a crystalline structure, nano-sized particles, a considerable amount of surface hydroxyl species, and a lowered band gap. The hierarchical heterostructure, including its form, specific composition, and functions, is essential in deciding the application of a sample. The MIL-100(Fe) was synthesized using a self-assembly method on TiO_2 nanosheets, resulting in the formation of hierarchical sandwich-like heterostructures. These structures were specifically engineered to break down methylene blue dye upon exposure to visible light with a wavelength exceeding 420 nm. This configuration improves the ability to efficiently capture, and separate electron-hole pairs created by light (Kambur et al., 2012) (Figure 5.4).

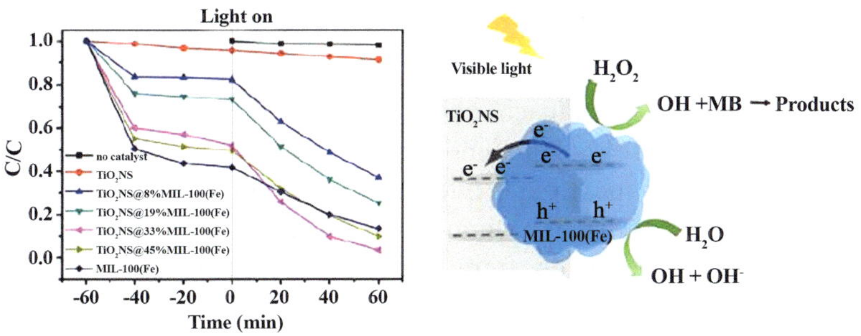

Figure 5.4. *Representation of hierarchical sandwich-like heterostructures for methylene blue degradation.*

Source: Meenakhsi Solanki, Creative Commons License.

Fluroxypyr (FLX) present in surface or groundwater can be effectively eliminated at alkaline pH levels, specifically between 8 and 10, through the application of Degussa P-25 Titania used as a photocatalyst. The reaction is governed by first-order kinetics. In 2006, Priya and Madras conducted a study on the first-order degradation of nitrobenzene and its derivatives using both combustion-generated and commercial TiO_2 catalysts while exposing them to UV radiation (Magdalane et al., 2016).

The conversion of aniline to azobenzene via photooxidation is accelerated when Fe_2O_3 is used as the photocatalyst in the solvent ethanol. The rate of this reaction is significantly increased when exposed to light with a wavelength of 254 nm. The photooxidation process, whether exposed to sunlight or UV light, is affected by the distinct properties of protic and aprotic solvents. Furthermore, not all photocatalysts demonstrate exceptional efficiency or effectiveness. Substances such as TiO_2, CuO, Fe_2O_3, ZnO, ZnS, PbO, ZrO_2, PbO_2, CdO, and HgO have the ability to degrade phenol when exposed to sunlight. However, chemicals such as Co_3O_4, MoO_3, CdS, Sb_2O_3, SnO_2, La_2O_3, Pb_2O_3, Sm_2O_3, Bi_2O_3, CeO_2, and Eu_2O_3 lack the capacity to breakdown phenol. In 2009, experiments revealed that tungsten oxide nanoparticles produced using the sol-gel technique showed efficient and complete breakdown of dyes upon exposure to laser radiation (Neppolian et al., 2007). Researchers investigated the impact of various factors, such as pH, inorganic salts, and H_2O_2, on the degradation of anionic dyes such as amido black, orange G, alizarin cyanine green, and Remazol brilliant blue R, employing ZrO_2 as a catalyst. The resulting samples underwent comprehensive analysis using a variety of analytical techniques, including FT-Raman spectroscopy, Fourier-transform infrared spectroscopy (FTIR), X-ray photoelectron spectroscopy, BET surface area analysis, UV-visible spectroscopy, and zero-point charge pH determination (Sotto et al., 2014).

The decontamination of Escherichia coli in water using nano-WO_3 was achieved through the sol-gel method coupled with laser irradiation. Metal oxides, including manganese oxides, ferric oxides, aluminum oxides, titanium oxides, cerium oxides, and magnesium oxides have demonstrated effectiveness in sequestering heavy metals from aqueous solutions. Their exceptional adsorption capacity for heavy metals is attributed to their large surface area and high affinity for these metals (Noureldine & Takanabe, 2016).

Researchers conducted a study in 2017 to investigate the degradation of organic pollutants, such as phenol, HCl, and rhodamine B dye. They employed a composite material composed of carbon nanodots and WO_3 nanorods. The degradation process was facilitated by visible light through a direct Z-scheme mechanism. The incorporation of carbon onto the photocatalyst surface resulted in a synergistic effect, leading to a threefold increase in performance compared to unmodified WO_3 nanorods (*refer to* Figure 5.5) (Suhaimi et al., 2022).

The combination of binary metal oxides alters the ability to harness solar energy, thus directly impacting its efficiency. Composites typically exhibit superior efficiency compared to their semiconductor counterparts. In 2006, Wu and colleagues synthesized CdS/TiO_2 composites via a solvothermal process assisted by microemulsion at a low temperature. The study demonstrated that the CdS semiconductor acts as a sensitizer for titania, thereby enhancing the degradation of methylene blue in water or nitric oxide in air when exposed to visible light (Edalati et al., 2020).

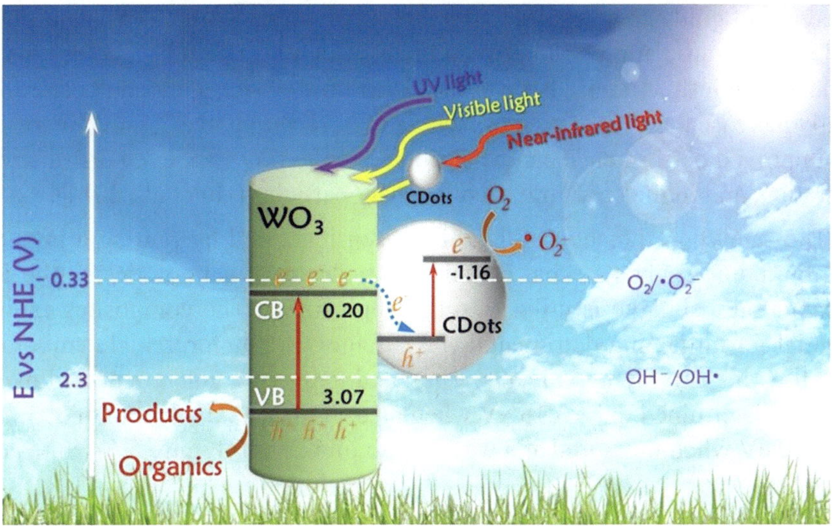

Figure 5.5. *Organic pollutants elimination via carbon nanodots/WO_3 nanorods composite.*

Source: Rakhsit Ameta, Creative Commons License.

The CdS/TiO_2 powder was prepared using a sol-gel method, with CdS concentration by weight ranging from 5% to 50%. The composite material displayed enhanced photocatalytic degradation of anionic azo-dye, specifically orange II, under illumination with wavelengths exceeding 400 nm. Notably, composites with lower CdS content exhibited superior performance compared to pure semiconductors. Various composite materials, such as ZrO_2–TiO_2, CuO–SnO_2, SnO_2/TiO_2, Fe_2O_3/SnO_2, TiO_2/WO_3, TiO_2–ZrO_2, and ZnO–SiO_2, have been used for the remediation of 4-chlorophenol, acid blue 62 dye, rhodamine-B, acid blue 62, phenol, methylene blue, and Hg (II) reduction (Tahir et al., 2020).

The TiO_2/SnO_2 and TiO_2/WO_3 systems show a rather modest rate of degradation of 4-chlorophenol at a wavelength of 369 nm. However, the pure element did not exhibit any deterioration at this particular wavelength. Nevertheless, when employing identical composites, the degradation rate of 4-chlorophenol increases at a wavelength of 435 nm. The impregnation approach was utilized to fabricate WO_3–TiO_2 photocatalysts. The interface created by the combination of monoclinic amorphous WO_3 and TiO_2 facilitates electron capture, hence enhancing the degradation of bisphenol A (Monfort & Petrisková, 2021). Nanocomposites can be formed by combining binary metal oxides or by using a suitable metal together with a semiconductor. Negatively charged gold nanoparticles are employed to alter the properties of TiO_2. The adherence of gold nanoparticles to the surface of a semiconductor is mainly due to the electrostatic attraction between them, which occurs at a pH close to the isoelectric point of titania. Au nanoparticles on the TiO_2 nanocomposite film greatly increase the degradation of bisphenol A, resulting in a 2.5-fold increase. The improved performance can be attributed to the tiny size, even distribution, separation, and high concentration of Au nanoparticles (Hu et al., 2014).

The catalytic efficiency of TiO_2 has been increased by putting a layer of TiO_2 on a silica substrate that contains embedded Ag nanoparticles. The ion implantation technique is utilized to accomplish this. The composites exhibit potential for mitigating detrimental and carcinogenic motor vehicle emissions as well as for employing solvents. The band gap of the interconnected metal oxides is determined to be 3.15 eV when measured at a wavelength of 393 nm, and 2.90 eV when measured at a wavelength of 427 nm (Figure 5.6) (Suresh et al., 2021).

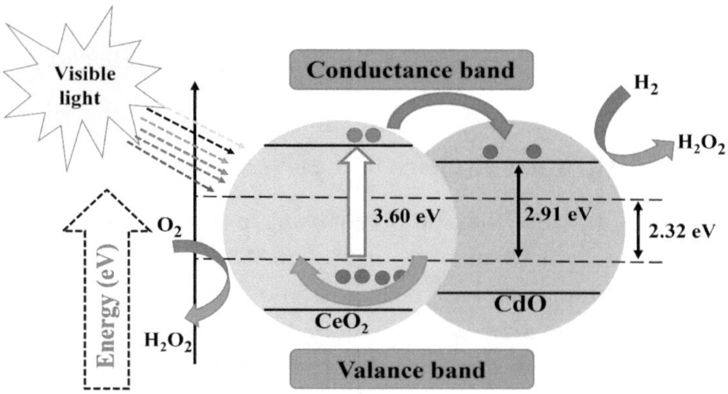

Figure 5.6. *Diagram of mechanism followed by CeO_2–CdO nanocomposites.*

Source: Suresh Ameta, Creative Commons License.

The addition of $Ti(x)Zr(1–x)O_2$ to polyether sulfone leads to the development of an ultrafiltration membrane. ZrO_2 hinders the development of macroscopic crystals in titania during the process. An observation was made that increasing the concentration of Zr leads to a proportional rise in the band gap, causing a shift toward the blue end of the spectrum (Panda et al., 2023).

The efficiency can be significantly improved by incorporating metal-free and co-catalysts. The $MnO.25CdO.75S/MoS_2$ composites were synthesized utilizing a one-step hydrothermal method. These composites have demonstrated efficacy in generating hydrogen through the process of water splitting when exposed to visible light. This enhanced efficacy can be ascribed to the existence of an interfacial stratum between MnO.25CdO.75S and MoS_2. The presence of this layer promotes the movement of electrons hence improving the separation of charge carriers (Figure 5.7) (Alzaid et al., 2023).

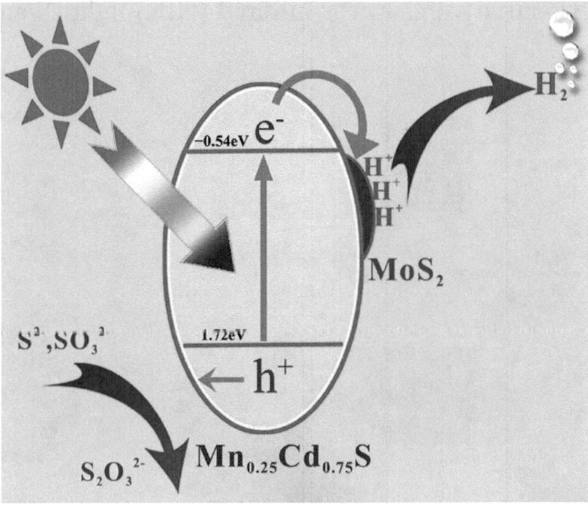

Figure 5.7. *Scheme of hydrogen production by using $Mn_{0.25}Cd_{0.75}S/MoS_2$ composites.*

Source: Surbhi Benjamin, Creative Commons License.

5.1.2. Ternary and Quaternary Oxides

$ZnFe_2O_4$, $SrFe_{12}O_{19}$, and $Sr_7Fe_{10}O_{22}$ were synthesized using both the polymeric approach and the solid-state method. These compounds were then used to facilitate the photodecomposition of isopropyl alcohol. The semiconductor Ba (In1/3Sn1/3M'1/3)O_3, where M denotes the metals Sn, Pb, Nb, or Ta, has a band gap ranging from 2.97 to 3.30 eV. Substituting non-transition metal cations with higher electronegativity enhances the sensitivity to a wide range of sunlight wavelengths (Sobahi & Amin, 2019).

Quaternary metal oxides such as Bi_4NbxTa1-xO_8I, $MnCoTiO_2$, and Pb_2FeNbO_6 have been manufactured using different techniques. These photocatalysts are capable of breaking down dye substances into biodegradable compounds. The ternary nanohybrid ZnS-Ag_2S-RGO composite, ZnO-RuO_2/RGO, and rGO/TiO_2/ZnO have demonstrated high efficiency in the photocatalytic oxidation of methylene blue and rhodamine B. The combination of ZnIn-MMO with g-C_3N_4 demonstrates superior performance in degrading rhodamine B compared to pure ZnIn-MMO and g-C_3N_4 alone, particularly under visible light conditions (Rajeshwar et al., 2018).

The Ag_2S/Ag_3PO_4 composites with a core-shell structure were synthesized by altering the Ag_2S composition. The composite with 5% Ag_2S has the maximum level of activity in visible light, while the composite with 50% Ag_2S demonstrates the highest activity under near-infrared (NIR) light (Wang et al., 2022) (Figure 5.8).

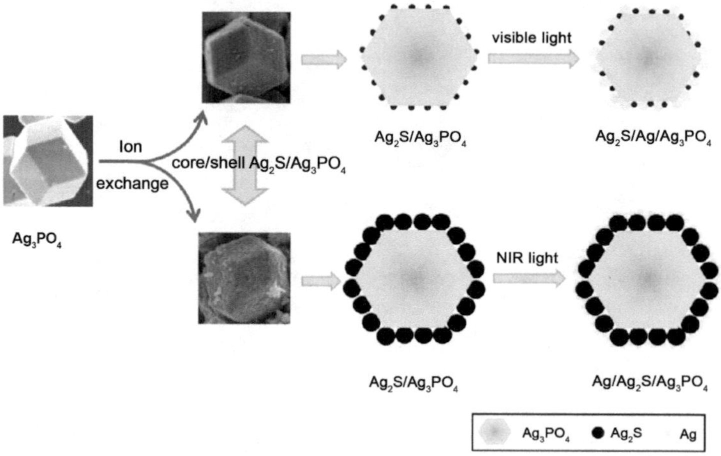

Figure 5.8. *Preparation scheme of core-shell Ag_2S/Ag_3PO_4 composites.*

Source: Meenakshi Solanki, Creative Commons License.

A one-pot method involving hydrolysis, precipitation, and UV photoreduction was employed to generate plasmonic Bi and BiOCl sheet adorned $ZnSn(OH)_6$. The resulting structure exhibited a cubic shape, with 14-faceted octa- and polyhedral systems. The alteration enhances the abstraction of photons, migration performance, and charge segregation, resulting in an 81-fold increase in the photooxidation of rhodamine B compared to pure $ZnSn(OH)_6$ (Marcolongo et al., 2022).

One experiment investigated the de-colorization of dye in the absence of light and catalyst. The researchers found that decolorization only occurred when both the catalyst and light source were present. There are several metal oxides, such as ZnO, TiO_2, SnO_2, WO_3, ZrO_2, ZnS, CdO, PbO, HgO, PbO_2, Fe_2O_3, $SrTiO_3$, $BiVO_4$, $PbCrO_4$, $PbWO_4$, $PbMoO_4$, etc., that demonstrate photocatalytic activity due to their low band gap, resilience, stability, affordability, versatility, and non-toxic nature. Various methods have been employed to synthesize different photocatalysts, including the sol-gel dip coating method, hydrothermal-assisted sol-gel technique, electrochemical deposition process, Stober method, precipitation method, combustion method, and others (Table 5.1) (Ao et al., 2018).

Table 5.1. *Band Gaps of Different Semiconductors*

Semiconductor	eV at 300K
ZnS (Zinc blende)	3.54
TiO_2	3.2
SnO_2	3.6
WO_3	2.6
ZnO	3.03
Fe_2O_3	2.2
CdS	2.42
Cu_2O	2.1
CdO	2.1
AlSb	1.58
CdSe	1.7
Wurtzite ZnS	3.91

Source: Meenakshi Solanki, Creative Commons License.

Most photocatalysts have a substantial band gap and are exclusively active in the ultraviolet zone. Sunlight comprises approximately 4% of UV light, with wavelengths shorter than 387 nm. Several studies have been conducted to expand the ability of semiconductors to respond to visible light and utilize 55% of the entire solar energy. Semiconductors offer viable options for addressing environmental contamination issues (Krishnan et al., 2024). Efforts have been undertaken to improve the photocatalytic efficiency of these materials by the application of diverse approaches. Semiconductors possessing a broad energy band gap and exhibiting a white hue do not absorb electromagnetic radiation within the visible region of the spectrum. Therefore, there is a requirement to alter these photocatalysts. Various techniques have been employed to alter the photocatalysts, such as modifying the surface, incorporating composites or coupler materials, introducing metallic and nonmetallic elements, and sensitizing the surface using organic colors and complexes of metals. These alterations can be achieved using a range of approaches such as metallic or nonmetallic doping, blending various metal oxides, codoping, sensitization, and so on (Siao et al., 2019).

5.2. MODIFICATIONS

Learning Objectives:

- Understand the doping and codoping techniques to enhance the photocatalytic activity.
- Comprehend the susbstitution and sensitization techniques for enhancement of photocatalytic activity.

Improving the performance of a semiconductor can be achieved through four fundamental methods. The mechanisms for creating localized states near the valence band include:

- The generation of localized states above the valence band;
- Utilization of semiconductors with low band gaps to generate localized states below the conduction band;
- Color center formation within the band gap;
- Modification of the surface (Liu et al., 2014).

Thus, the methods employed in alterations include:

- Introducing metal and/or nonmetal into a material to enhance its properties by doping.
- Simultaneously doping a material with both donor and recipient materials in different combinations.
- Combining different photocatalysts to create a composite material.
- Enhancing the light absorption capabilities of a material through sensitization.
- Replacing certain atoms in a material with others to modify its properties.
- Other methods or techniques not falling into the previous categories.

Large band gap photocatalysts exhibit a lack of color and have a strong tendency for electron-hole recombination (He et al., 2020) (Figure 5.9).

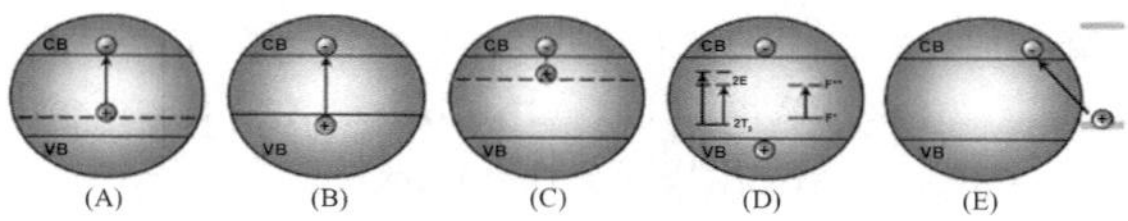

Figure 5.9. *Illustration of (A) localized states above valence band; (B) semiconductor with narrowed Eg; (C) localized states below conduction band; (D) color centers formed in the Eg; and (E) surface modification.*

Source: Ghorban Asgari Creative Commons License.

5.2.1. Doping

Doping represents the deliberate modification of a photocatalyst through the introduction of impurities into a semiconductor material, thereby diminishing the energy differential between its valence and conduction bands. This technique enables the incorporation of both metallic and non-metallic elements into the semiconductor lattice. Each type of dopant exerts a distinct influence on the crystalline framework of the semiconductor (Khairy & Zakaria, 2014). The integration of metal and non-metal constituents enriches the photocatalyst's responsiveness to visible light by generating supplementary energy levels, denoted as impurity states, positioned between the conduction bands and valence (*see* Figures 5.10 and 5.11). The introduction of these additional energy levels serves to narrow the energy gap separating the valence and conduction bands, facilitating the transition of excited-state electrons to the conduction band. Doping is integral to the practice of band gap engineering, which involves the deliberate manipulation of electron or hole states within the photocatalytic semiconductor (Huang et al., 2016).

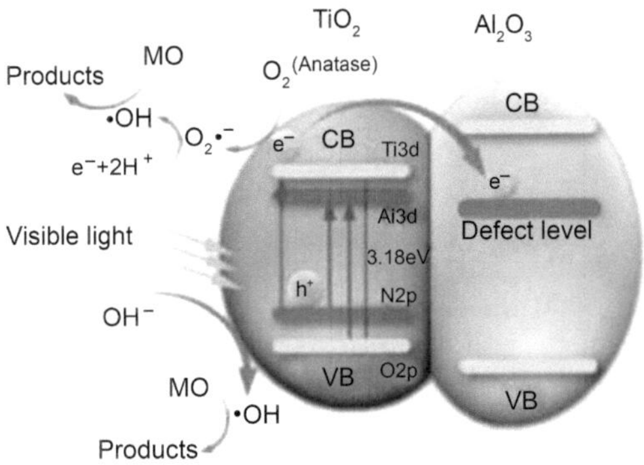

Figure 5.10. *Metal doping effect on conduction band.*

Source: Tarek Balchi, Creative Commons License.

Metallic dopants, including Co, Cr, W, Cu, Zn, Mo, Mn, Fe, and Ni, induce the formation of a novel band situated below the conduction band. Conversely, non-metal elements such as nitrogen, phosphorus, silicon, sulfur, chlorine, selenium, bromine, and iodine engender the creation of a higher energy band positioned above the valence band. This strategic incorporation serves to mitigate recombination events by bolstering the electron capture process, thereby culminating in an elevation of photocatalytic activity. The inclusion of metal dopants not only enhances the structural integrity and surface area but also augments the light utilization efficiency for catalytic processes, electrical characteristics, and electromagnetic properties of photocatalytic semiconductors (Jiang et al., 2017).

There are three possible scenarios in which nonmetallic compounds can undergo a reaction with the oxides of a photocatalyst. At first, the dopant can combine with the photocatalyst's oxide to create a hybrid structure. Furthermore, an oxygen site can be replaced with a dopant. Then the dopant can be introduced into an oxygen-deficient location and O_2 sub-stoichiometry, acting as a hindrance to the process of reoxidation. The formation of oxygen vacancies was witnessed due to the addition of a nonmetallic carbon dopant. This led to the emergence of Ti_3I species positioned between the valence band and conduction band in the band structure of TiO_2. The size of dopant particles plays a critical role in influencing how oxygen is replaced in a photocatalyst. Smaller atoms like C, N, or F can take the place of the O atom on the TiO_2 surface, whereas larger atoms like phosphorus, sulfur, chlorine, and bromine can replace the O on the SrO-terminated surface (Kudo & Sekizawa, 2000).

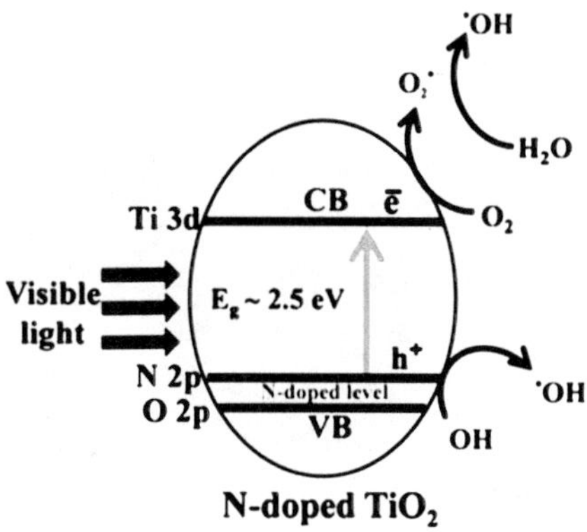

Figure 5.11. *Effect of nonmetal (nitrogen) doping on valence band.*

Source: Yuan Li, Creative Commons License.

Similar energy levels characterize photocatalyst oxides and nonmetal dopants. The catalyst's band gap reduces, making it better at absorbing visible wavelengths. Band gap modification alters a semiconductor's morphological, structural, electrical, and optical properties, which affects light absorption, charge carrier mobility, surface area, and redox potential. Donor and receiver species affect the photocatalyst's electrical structure and reactivity. Doping lowers the photocatalyst's band gap, improving CB and VB dispersion. Red-shift occurs when the system is modified, shifting the absorption spectra edge to longer wavelengths. This bathochromic shift reduces the band gap or introduces an intra-band gap state through doping, capturing more photons (Ali et al., 2013).

5.2.2. Codoping

Combining different donor and acceptor species can be used to decrease the band gap in a process known as codoping. One approach to achieve this is by increasing the valence band edge and decreasing the conduction band edge. It enables the achievement of activity across a broad spectrum of solar wavelengths. There have been cases where both metallic and nonmetallic materials have been introduced together, as shown in Figure 5.12 (Nur et al., 2022).

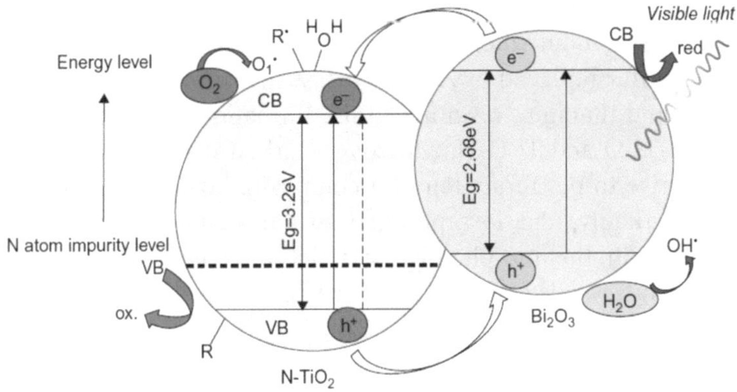

Figure 5.12. *Effect of codoping.*

Source: Yuan Li, Creative Commons License.

In their 2013 study, Yan et al. observed that this approach effectively addresses challenges such as carrier recombination, solubility limitations, restricted carrier mobility, and diminished responsiveness to visible light within the host material. Through the codoping of non-metallic and metallic elements such as N-W, C-Mo, N-Nb, Ce-F, and N-Ta, enhanced efficiency was achieved without altering the position of the conduction band edges. The presence of both Ce and F elements exerted an influence on the internal structure and optical band gap of Bi_2WO_6. Furthermore, the combination of a metal (Sm) and a non-metal (N) led to reduced crystal growth, minimized charge carrier recombination, and facilitated the transformation from the anatase to rutile phase.

Codoping Ga and Cu with ZnS via hydrothermal synthesis resulted in a material with exceptional stability, even under prolonged irradiation. Employing the hydrogel method, researchers succeeded in fabricating N/F/anatase titania featuring {001} facets, thereby enhancing performance by augmenting the driving force for electron transfer. This improvement was attributed to the enhanced efficiency of electron injection from the dye's lowest unoccupied molecular orbital (LUMO) into the semiconductor's conduction spectrum (Xu et al., 2008).

The distinctive composition of various materials incorporated into titania enables it to exhibit responsiveness to a broad spectrum of light, encompassing ultraviolet radiation, visible wavelengths, and direct sunlight. This inherent property renders it highly efficacious for the removal of diverse organic compounds. In their 2009 investigation, Wang and colleagues employed l-cysteine, a biomolecule rich in carbon, nitrogen, and sulfur, as a versatile precursor. Through the hydrothermal method, they successfully synthesized C-N-S-tridoped titanium dioxide nanocrystals. Through strategic elemental substitution, sulfur replaced oxygen atoms, while nitrogen formed bonds with both oxygen and titanium, resulting in the formation of distinctive structures denoted as N-Ti-O and Ti-O-N. Moreover, carbon deposition onto the titania surface gave rise to the formation of a composite layer comprising carbonate species. Remarkably, the incorporation of l-cysteine exerted a substantial influence on both the morphology and the final crystal structures of the synthesized nanocrystals (Jaiswal et al., 2016).

5.2.3. Coupled Semiconductors or Composites

Creating composites by combining semiconductor components is a powerful method to improve the ability of semiconductors to respond to light in the visible range of the electromagnetic spectrum. When choosing semiconductors for composite preparation, it is important to have semiconductors with band gaps that can be adjusted (Zhang et al., 2017).

Various combinations of photocatalysts such as Ag_3PO_4, $FeTiO_3$, $W_{18}O_{49}$, CdSe, CdS, ZnO, $NiTiO_3$, $CoTiO_3$, Fe_2O_3, and SnO_2 have been paired with TiO_2. Composite materials have enhanced photocatalytic activity compared to individual components, due to the synergy effect. According to reports, the combination of Cr_2O_3 and SnO_2 has been discovered to demonstrate excellent photocatalytic performance when exposed to visible light. This is attributed to the presence of well-formed nanoparticles, lesser crystal size, and a heightened sensitivity to visible light (Han et al., 2014).

The sample exhibits a degradation rate of 98% for rhodamine B within a 60-minute timeframe when exposed to visible light irradiation. Doping linked photocatalysts with different combinations of nonmetals and metals can further enhance their photocatalytic activity. Doping results in the formation of a homogeneous dispersion of particles with minimal clumping, and a porous microstructure. The methyl orange undergoes active breakdown as a result of a decline in charge recombination. Coupling can also entail the alteration of the outside layer of mesoporous SnO_2 through the addition of other metal oxides, such as TiO_2 or Al_2O_3 (Keller & Garin, 2003).

Remember

Several factors influence the efficiency of photocatalysis, such as the choice of photocatalystx, light intensity, pH levels, temperature, and the presence of co-existing ions.

5.2.4. Substitution

The substitution of Ti_{41} with Fe_{41} in a $CaFe_2O_4$–$MgFe_2O_4$ bulk hetero-junction lattice photocatalyst resulted in an observed enhancement in photocurrent. The sample with Ti ions showed a greater quantum yield for the photodecomposition of the H_2O-CH_3OH combination, which is likely due to the efficient separation of charges (Ashwin Kishore & Ravindran, 2017). Wang and his team also investigated the impact of substituting W metal in WO_3 with various metals, such as Mo, Ti, Zr, and Hf, on the alteration of the band gap. Based on their findings, when W in WO_3 is replaced with another metal of the same valency, like Chromium and Molybdenum, the band gap decreases as the CB edge moves towards the lower side.

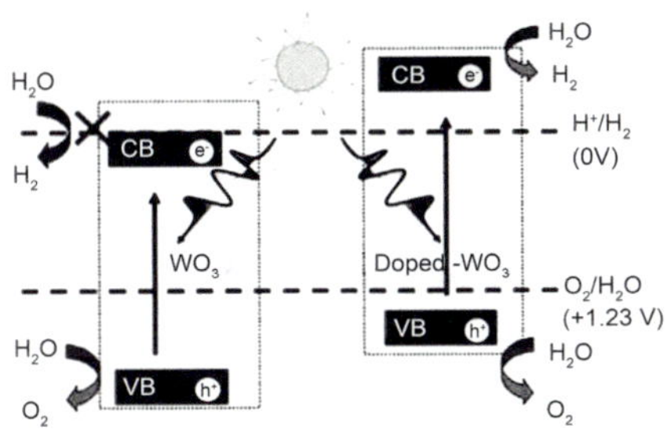

Figure 5.13. *Electron transfer mechanism in WO_3 and doped WO_3.*

Source: Ghorban Asgari, Creative Commons License.

When metals with lower valency, like Zr, Ti, and Hf, are added, the band gap increases because the conduction band edge moves higher. When oxygen is replaced with sulfur, there is a decrease in the band gap. This phenomenon occurs due to the formation of localized states within the valence band (VB), leading to an upward shift of the conduction band (CB) minimum. These effects are shown in Figure 5.13 (Yuan et al., 2007).

Identical dopant, in either anatase or rutile form, exhibits distinct electrical properties. Therefore, several dopants can produce an equivalent impact on the band gap. For example, rutile TiO_2 that is doped with Si, Ge, Sn, and Pb shows a drop in band gap of around 0.1 eV, with a value of about 0.55 eV (Zhao et al., 2020).

5.2.5. Sensitization

Using sensitization as an alternative method, there is potential for modifying the surface of a photocatalyst. Photocatalytic systems, such as solar cells, can make use of dyes or complexes due to their redox characteristics and their ability to respond to visible light. When exposed to sunlight, dyes can transfer electrons to the conducting band of the semiconductor, thus initiating a catalytic reaction (Figure 5.14) (Chen & Wang, 2014).

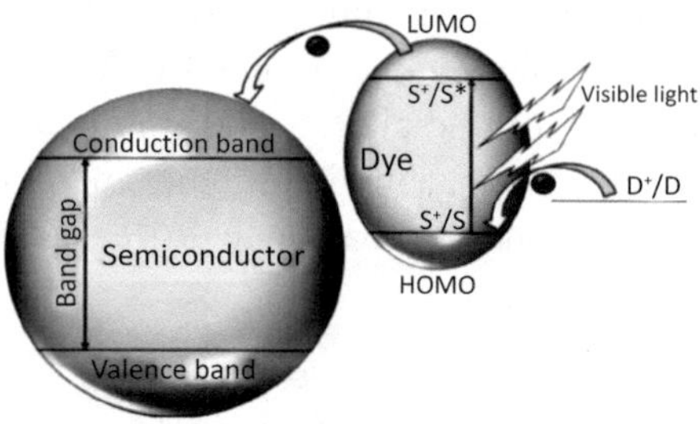

Figure 5.14. *Role of a sensitizer in photocatalysis.*

Source: Tarek Balchi, Creative Commons License.

Substances, whether organic or inorganic, can attach themselves to the surface of a semiconductor through chemisorption or physisorption. Applying sensitizers to the surface of a semiconductor is part of the sensitization process. They can be classified as (Vaz & Pérez-Lorenzo, 2023):

- Artificial/Synthetic photosensitizer;
- Natural photosensitizers;
- Autogenous sensitizer.

Commercially available dyes like Ru complexes, porphyrin dye, Ru-polypyridyl-complex sensitizers, and quinoxaline-based organic sensitizers are referred to as synthetic sensitizers (Reginato et al., 2020). The efficiency of photocatalysis is influenced by the reactivity of the photocatalyst towards light radiation. Using certain chemical compounds that contain chromophores, including natural pigments or dyes, can enhance the light sensitivity of the semiconductor. These chemicals are commonly referred to as photosensitizers. A photosensitizer has the remarkable ability to absorb light and then transfer that energy to the semiconductor. With their redox characteristics and sensitivity to visible light, dyes are well-suited for use in photocatalytic devices and solar cells. Just by exposing dyes to visible light, they have the ability to transfer electrons to the conducting band of the semiconductor, resulting in the initiation of a catalytic reaction (Ohno et al., 2006).

Various natural substances, including those derived from plants, fruits, flowers, and even animals and insects, can serve as sensitizers. These substances encompass a wide range of compounds such as polyphenols, carotenoids, chlorophyll and its byproducts, and anthocyanin, which can be obtained from sources like Pastinaca sativa, Monascus red, and Beta vulgaris.

The performance depends on the specific characteristics of the natural pigment. Natural dyes are highly preferred over synthetic dyes because of their plentiful raw materials, simple extraction process, affordability, and environmentally friendly characteristics (Toumazatou et al., 2017).

The naturally occurring organic dyes have the potential to replace certain synthetic dyes that are currently used as sensitizers. Natural pigments have unique structural and optical characteristics that make them effective photosensitizers. Kushwaha et al. in their study acquired extracts from various plant sources, including teak, eucalyptus, tamarind, and the flower of crimson bottle brush. These extracts were used to enhance the efficiency of dye-based solar cells in capturing visible light.

Synthetic dyes are more effective than natural dyes. Improved effectiveness depends on various factors:

- Technique for isolating dye;
- Acidity level of the solvent employed for isolation; and
- Influence of various solvent components on the absorption spectra (Fox, 1983).

The choice of water and ethanol as the extracting solvent, along with the acidity of the dye solution, can have an impact on the reliability and optical properties. Natural dye boosts the efficiency of light absorption and promotes sustainability, while also increasing the rate of electron transfer. In their study, Benjamin et al. (2011) utilized natural pigment extracts like chlorophyll and anthocyanin to sensitize ZnO. They then used this combination to facilitate the photocatalytic decomposition of dyes. Through this discovery, it was found that the natural pigment can sensitize the photocatalyst, leading to a reduction in its band gap. As a result, the photocatalyst becomes more effective in utilizing visible and infrared rays (Xiao et al., 2008).

Flowering plants in the Caryophyllales order have pigments called Betalains that are rich in nitrogen. These pigments have excellent light absorption properties and also act as antioxidants. These are linked to different co-pigments, which play a role in altering light absorption characteristics. Pigment carboxylic functional groups readily adhere to the surface of photocatalysts such as TiO_2. The pigment derived from red rose flower petals has the potential to serve as a sensitizer in various dye-sensitized solar cells. In addition, it serves as an antioxidant, safeguarding the fat found in fermented dairy products like yogurt.

The sorghum red pigment, derived from the shell of sorghum, displays its peak absorption wavelength. The pH value has an impact on the durability of this pigment when exposed to light. The *Saraca indica* leaf extract serves as a reducing agent in the production of CuO nanoparticles. This extract functions as a stabilizing agent for the catalyst. Utilizing extracts of *S. indica* in the manufacturing process allows for the determination of the size and morphology of CuO nanoparticles, employing a green approach (Li et al., 2002).

Similar to other sensitizers, natural dye molecules are susceptible to degradation when exposed to light. When it breaks down, it completely removes all traces of organic compounds from the solution. When the sensitizer deteriorates, it causes the underlying catalyst to become deactivated during the recovery process.

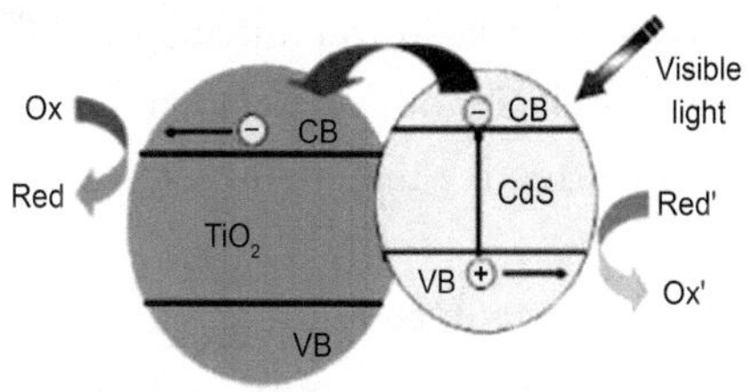

Figure 5.15. *Representation of charge transfer between CdS and TiO_2.*

Source: Violeta Vadillo, Creative Commons License.

However, synthetic dyes are more efficient than natural ones. Producing synthetic dyes can be quite challenging. These substances are costly, toxic, and result in a reduced output. On the other hand, the process of extracting natural dyes is straightforward, cost-effective, adaptable to various temperatures, completely biodegradable, and poses no hazards. Unfortunately, most sensitizers tend to degrade or corrode when they come into contact with water-based solutions.

Therefore, they are unsuitable for utilization in photocatalytic processes for water treatment. Various types of synthetic photosensitizers include plasmonic metal nanostructures, quantum dots, and carbon nanostructures. These are combined with transition metal oxides with wide-band gaps to create highly efficient photocatalysts that can be activated by visible light. This concept is depicted in Figure 5.15 (Gonuguntla et al., 2023).

5.2.6. Miscellaneous

Titania, characterized by its exceptional purity and a band gap exceeding 3.1 eV, stands out for its pronounced reactivity within the ultraviolet (UV) spectrum. This inherent property positions it as promising for the production of clean energy and the resolution of environmental challenges. In a significant study conducted in 2015, Yaghoubi et al. observed a noteworthy alteration in the energy gap of titania, attributable to the emergence of midgap states during the synthesis process, which utilized a solution-phase approach. These resulting mixed-phase TiO_2 nanoparticles exhibited a band gap of approximately 2.2 eV, along with the presence of oxygen vacancies and a notably high surface area, features that substantially bolster their photoreactivity (Sheng et al., 2023).

Upon absorption of specific wavelengths of light, an electron within the valence band undergoes excitation, transitioning to the conduction band. This transition occurs across the band gap, which denotes the energy difference between the valence and conduction bands and represents the minimum energy required to stimulate electron excitation. When a photon possessing adequate energy is absorbed by the photocatalyst, it can initiate either reduction or oxidation processes through photocatalysis. Following excitation, electron-hole pairs are generated on the surface of the photocatalyst, with the electron occupying the conduction band and the hole remaining in the valence band (Dake et al., 2020).

In one scenario, the charge carriers recombine, resulting in the generation of heat. Alternatively, in another scenario, the charge carriers may participate in additional reactions with electron donors or acceptors present on the surface of the photocatalyst. In this latter scenario, an electron or hole may interact with dissolved water or oxygen, leading to the formation of O_2^-, HOO•, or OH• free radicals. These highly reactive oxidizing species, in conjunction with other compounds such as H_2O_2 and O_2, play a pivotal role in driving the photooxidation reaction, thereby facilitating the degradation of contaminants and the advancement of environmental remediation efforts (Fenoll et al., 2012).

Practice Problem

Discuss the applications of photocatalysis in wastewater treatment and provide examples of real-world scenarios where photocatalysts have been successfully utilized.

Solution to Practice Problem

Photocatalysis has diverse applications in wastewater treatment, including the degradation of organic pollutants, removal of heavy metals, antimicrobial activity, and oil degradation. For example, semiconductor photocatalysts like titanium dioxide (TiO_2) have been used to degrade pharmaceutical residues in pharmaceutical wastewater, while coupled semiconductors have been employed to remove heavy metals like lead and cadmium from industrial effluents. Additionally, photocatalysis has been applied in municipal wastewater treatment plants to disinfect water and eliminate harmful pathogens, improving water quality and public health.

5.3. WASTEWATER TREATMENT

Learning Objectives:

- Comprehend the applications of photocatalysis in wastewater treatment such as organic pollutants elimination, removal of heavy metal and degradation of oil.

Effluents originating from many sources such as households, municipalities, industries, sludge, etc., contain a multitude of nonbiodegradable organic matter, xenobiotic compounds, crude oil, petrochemical waste, colors, and other substances. Industrialization has significantly contributed to pollution. Various businesses, such as textile, chemical, paper products, pesticides, and fertilizer industries, release their waste into adjacent water bodies without undergoing any initial treatment. This is a significant factor contributing to water pollution. Water pollution is becoming a significant peril to the ecosystem (Borges et al., 2016). Numerous conventional techniques, including adsorption, coagulation, sedimentation, chlorination, reverse osmosis, ozonation, filtration, trickling filter, UVC disinfection, activated sludge process, aerated lagoon, oxidation ditch, and stabilization ponds, have been employed extensively over the years. Despite their widespread use, these methods often fall short when confronted with the challenge of treating large-scale wastewater streams. Their limitations primarily stem from the considerable costs associated with capital investments, implementation, maintenance, and land area requirements, rendering them inadequate or impractical for addressing the needs of expansive wastewater treatment systems. Furthermore, these technologies do not sufficiently reduce the pollutants in the effluents to a level where they may be recycled, and they also necessitate several days for examination. Consequently, it is crucial to create new purifying methods that are hygienically friendly. Refinery effluent contains varying levels of aliphatic and aromatic hydrocarbons from petroleum, which are both hazardous and resistant to degradation through conventional treatment procedures. Researchers used titania nanoparticles as a photocatalyst in the UV/TiO_2 method to remediate the wastewater from petroleum refineries (Sinar Mashuri et al., 2020). Researchers have advocated for the use of photocatalytic decomposition process to break down and eliminate difficult-to-biodegrade inorganic as well as organic pollutants in industrial wastewater. This approach is employed to prevent the release of toxic or nonhazardous substances into mainstream water bodies. This technique produces extremely reactive hydroxyl radicals that oxidize the substances in the solution and fully convert them into water, CO_2, and organic substances. An advanced oxidation process (AOP) utilizes nanostructured oxides of metal or sulfide semiconductors like TiO_2, SnO_2, ZnO, SiO_2, ZnS, CeO_2, MnS, CuS, CdS, $SrTiO_3$, $BaTiO_3$, hematite (α-Fe_2O_3), $PbTiO_3$, Bi_2WO_6, and others. The advanced Oxidation Process is a very efficient approach used to oxidize stubborn contaminants, specifically phenolic chemicals. When photocatalysts and UV light are used together, they can convert organic pollutants or dangerous organic molecules into substances that are not poisonous or less damaging, such as carbon dioxide and water. In addition, it has the capability to eradicate specific microorganisms such as bacteria and some viruses during the secondary water treatment process. AOPs have demonstrated their cost-effectiveness in converting dangerous compounds or contaminants, even when present in very small quantities, into harmless components. Additionally, it prevents the adverse effects of contaminants that possess high toxicity, and persistence and are challenging to remediate (Amin et al., 2014).

5.3.1. Antimicrobial Activity

Wastewater may potentially harbor pathogenic bacteria in addition to organic contaminants. Photocatalysts facilitate the eradication of microorganisms such as bacteria, viruses, and fungi, and they can even break down the cellular structure itself. Cell death results in the release of toxins. The poison that is emitted also undergoes decomposition by photocatalytic action. In 2017,

Sangchay developed a coating of WO_3-doped TiO_2 on charcoal using the microwave-assisted sol-gel technique. Enhanced photocatalytic activity was demonstrated in the breakdown of a methylene blue solution, as well as against *E. coli*. The addition of 1% WO_3 to TiO_2 covered with activated charcoal significantly enhanced its antibacterial properties in comparison to pure titania. The Giardia lamblia-infested aqueous solution was sterilized using titania in the presence of UV irradiation. The adherence of a photocatalyst to the surface of a substrate aids in the process of photocatalytic degradation of acid orange 7, in conjunction with microorganisms such as the bacterium *E. coli* and viruses like Herpes simplex (Chu et al., 2021).

5.3.2. Organic Pollutants Elimination

The photocatalyst is not the sole determinant for the elimination or breakdown of aqueous contaminants. When molecular oxygen, which acts as an electron acceptor, is present, phenol red is only 60% bleached. However, in a concentration of hydrogen peroxide, it experiences more than 90% deterioration when exposed to visible light. The chemical breakdown of organic pollutants is influenced by temperature, which is an essential aspect. Oxidants have a crucial role in the breakdown of organic pollutants. In a study, a reduction of 18.26% was reported in phenol concentration while utilizing iron-modified titanium dioxide, namely Fe_2O_3/TiO_2. Additionally, a breakdown rate of 27.87% was observed when the appropriate amount of H_2O_2 was utilized. The elimination of substituted phenols has been investigated by the utilization of metal oxides such as TiO_2 or ZnO. However, the primary challenge is in the separation of the catalysts from the solution. Cobalt ferrite ($CoFe_2O_4$) is a semiconductor that exhibits both magnetic properties and a narrow band gap. As a result, it may be readily separated from a solution by utilizing a magnetic field from the outside. A study reported a significant 96% reduction in the concentration of the water contaminant 2-nitrophenol through the utilization of Cu slag-supported $CoFe_2O_4$ (Rashed, 2013). Under the influence of visible light, N-TiO_2 facilitates the dechlorination process of 2,4-dichlorophenol, resulting in the release of chlorine in the form of chloride ions. The process of fully removing chlorine takes 2 hours when using a catalyst that has been modified, while it takes approximately 3 hours when using pure ZnO. Furfural is an organic substance utilized as a solvent in petrochemical setups, petroleum refineries, and other manufacturing facilities for the purpose of separating hydrocarbons. It is nonbiodegradable, extremely durable, and undergoes a gradual process of breakdown. Consequently, it serves as a significant contributor to water contamination. The presence of furfural in wastewater has significant detrimental effects on living organisms. The substance is absorbed by the skin and has detrimental effects on the neurological system, causing damage to cell membranes and hindering metabolism. Due to its hazardous nature, it is necessary to degrade, mineralize, and remove it from the effluent. Granular activated carbon (AC) is characterized by its porosity and low density. A Titania catalyst based on granular activated carbon was produced using a straightforward sol-gel approach. The catalyst has a total pore volume of 0.13 cm^3/g and a specific surface area of 35.91 m^2/g. The study demonstrated that the furfural degradation rate reached approximately 95% within a time frame of 80 minutes. Furthermore, the catalyst employed in the study exhibited the ability to be reused for at least four cycles, with just a little decrease of 2% in furfural removal effectiveness. The composite has a lower total pore volume compared to the GAC supporter. As a result, the titania material is fully enclosed within the pores of the supporter (Melvin & Leusch, 2016).

5.3.3. Removal of Heavy Metal

Samadi et al. (2014) documented the process of fixing Cu-titania/chitosan onto a polycarbonate substrate. The stabilized thin film was synthesized using a straightforward sol-gel process, making it dependable, reproducible, and cost-effective. It demonstrated exceptional efficacy in eliminating heavy metals, including Pb^{2+} and Cr^{6+}, from water-based solutions. Therefore, the immobilization of a catalyst through several methods offers a resolution to issues related to the segregation, recycling, and deactivation of the catalyst. The TiO_2 and maghemite (γ-Fe_2O_3) were synthesized using coprecipitation and hydrothermal methods. Subsequently, these substances were combined in ratios of 1:10, 1:20, 1:60, 1:80, and 1:100 and coated with polyvinyl alcohol (PVA)-alginate beads. The encapsulated composite effectively eliminates 100% of the hazardous element, Pb(II), from an aqueous solution within 100 minutes under neutral conditions and sunshine exposure (Qasem et al., 2021).

5.3.4. Degradation of Oil in Wastewater

Oil spills are a significant, often overlooked, source of nonbiodegradable pollutants that contaminate water and disrupt entire aquatic ecosystems. Photocatalytic techniques can efficiently destroy crude oil or other hydrocarbon contaminants that are discharged from oil refineries and petroleum. This technology is both cost-effective and efficient in converting pollutants in water, such as carbon dioxide and biodegradable organics. Therefore, it has extensive uses in mitigating oil pollution in coastal areas, open waters, refineries, and water treatment facilities. Crude oil components in polluted water do not break down by exposure to light, but instead undergo degradation through heterogeneous photocatalytic reactions. The level of degradation achieved is approximately 90% (Lan et al., 2009). The anatase form of Titania has been demonstrated to be the most effective for oil spill remediation. The removal of polycyclic aromatic hydrocarbons (PAH) and total organic carbon (TOC) from oil sludge was achieved by heterogeneous photocatalysis. This process involved the use of titanium dioxide in the presence of hydrogen peroxide and was carried out under sunlight. Polycyclic aromatic hydrocarbons are fully removed within 96 hours, and there is a 91% decrease in total organic carbon within 144 hours using 5 milligrams of TiO_2. The addition of H_2O_2 in Photo-Fenton systems, both homogeneous and heterogeneous, enhances the treatment efficiency of oily wastewater compared to systems without Fenton. Photocatalysis decomposes the hydrocarbon spill into oil, resulting in the formation of solely biodegradable chemicals. Toluene is the aromatic component of the oil spill. It can be broken down by a thin coating of N-doped TiO_2 nanotubes that is applied to a levee at the sea surface level. The precipitation approach was used to generate maghemite nanoparticles (γ-Fe_2O_3), which were found to eliminate 90% of toluene from seawater after 2 hours under sunshine (Mustafa et al., 2014).

Remember

Photocatalysis holds promise for decentralized water treatment systems, offering a sustainable and energy-efficient solution for small-scale wastewater treatment applications.

5.4. APPLICATIONS OF PHOTOCATALYSIS

Learning Objectives:

- Understand the properties of photocatalysts
- Comprehend the applications of photocatalysis.

Photocatalysts are highly versatile materials due to their appealing features, including optical transmittance, homogeneity, low resistance, structural hardness, thermal stability, hydrophobicity, and piezoelectric activity. As a result, they are well-suited for various applications such as gas sensors, Li-ion batteries, photovoltaic systems, supercapacitors, light-emitting diodes, display devices, solar cells, DSSCs for electrochemical water splitting, antibacterial purposes, antifogging applications, self-cleaning properties, and more.

Photocatalysis is a prominent area of green chemistry that relies on the utilization of light, which is readily accessible and does not contribute to atmospheric pollution. Wastewater treatment involves several chemical and physical processes. Among these, photocatalytic treatment has emerged as a highly effective and environmentally acceptable technology for removing organic contaminants from wastewater. Alternative techniques often generate additional contaminants or discharge toxic substances into the environment. In contrast, photocatalytic treatment of wastewater represents a significant advancement in promoting environmentally friendly chemistry and is considered a crucial technology within the realm of advanced oxidation processes. In the near future, photocatalysis will demonstrate its value and establish a prominent role in the future treatment of wastewater (Agrios & Pichat, 2005).

SUMMARY

- This chapter explores the principles, uses, and progress of photocatalysis technology. Photocatalysis is a process that utilizes semiconductor photocatalysts to trigger oxidation and reduction processes when exposed to light. This process ultimately leads to the breakdown of pollutants in wastewater.
- The chapter commences with an examination of several semiconductor photocatalysts, such as TiO_2, ZnO, and other metal oxides. It emphasizes their distinct characteristics and appropriateness for diverse applications.
- The detail provides a discussion of several modification strategies, including doping, codoping, linked semiconductors, substitution, and sensitization. It explains how these techniques effectively improve thc photocatalytic properties of semiconductor photocatalysts.
- This chapter discusses numerous uses of photocatalysis in wastewater treatment, including the breakdown of organic contaminants, elimination of heavy metals, antibacterial properties, and oil decomposition. Photocatalysis has proven to be beneficial in tackling pollution concerns in many municipal and industrial waste streams, as demonstrated by practical instances and case studies.
- The chapter offers a thorough and inclusive examination of photocatalysis for the treatment of wastewater, emphasizing its adaptability, efficiency, and capacity to tackle pollution issues and advance sustainable development in water management.

REVIEW QUESTIONS

1. Describe the various modification techniques used to enhance the photocatalytic activity of semiconductors for wastewater treatment.
2. Explain the principles underlying photocatalysis and how semiconductor photocatalysts utilize solar energy to initiate oxidation and reduction reactions in wastewater treatment.
3. Compare and contrast binary oxides, ternary oxides, and quaternary oxides as photocatalysts for wastewater treatment.
4. Explore the applications of photocatalysis in wastewater treatment, including the removal of organic pollutants, antimicrobial activity, removal of heavy metals, and degradation of oil. Provide examples of real-world applications in each category.

5. Discuss the role of sensitization in enhancing the photocatalytic activity of semiconductor photocatalysts. How does sensitization improve the absorption of visible light and promote more efficient pollutant degradation?

MULTIPLE CHOICE QUESTIONS

1. Which of the following is not a modification technique employed to enhance photocatalytic activity?

a. Doping
b. Codoping
c. Coupled semiconductors
d. Biodegradation

2. Photocatalysis for wastewater treatment primarily relies on the utilization of:

a. Thermal energy
b. Solar energy
c. Mechanical energy
d. Electrical energy

3. What is the primary function of sensitization in photocatalysis?

a. Enhancing the antimicrobial activity
b. Increasing the absorption of visible light
c. Reducing heavy metal content
d. Improving oil degradation efficiency

4. Which type of oxides are commonly used as photocatalysts?

a. Binary oxides
b. Acidic oxides
c. Organic oxides
d. Noble oxides

5. What is the main advantage of coupled semiconductors in photocatalysis?

a. Increased stability
b. Enhanced photocatalytic activity
c. Reduced cost
d. Improved selectivity

Answers to Multiple Questions

1. (d); 2. (b); 3. (b); 4. (a); 5. (b)

REFERENCES

1. Agrios, A. G., & Pichat, P. (2005). State of the art and perspectives on materials and applications of photocatalysis over TiO_2. *Journal of Applied Electrochemistry, 35*, 655–663. https://doi.org/10.1007/s10800-005-1801-6.
2. Ahmed, S., Rasul, M. G., Martens, W. N., Brown, R., & Hashib, M. A. (2010). Heterogeneous photocatalytic degradation of phenols in wastewater: A review on current status and developments. *Desalination, 261*(1–2), 3–18. https://doi.org/10.1016/j.desal.2010.01.024.
3. Ali, S. M., Hussain, S. T., Bakar, S. A., Muhammad, J., & ur Rehman, N. (2013). Effect of doping on the structural and optical properties of SnO_2 thin films fabricated by aerosol assisted chemical vapor deposition. In *Journal of Physics: Conference Series* (Vol. 439, pp. 12013). IOP Publishing. https://doi.org/10.1088/1742-6596/439/1/012013.
4. Alzaid, M., Mohamed, W. S., Alanazi, R., Alsohaimi, I. H., Hassan, H. M., Hadia, N. M. A., & Abu-Dief, A. M. (2023). Novel (Y_2O_3) -(CdO) binary mixed oxide nanocomposites: Facile synthesis, characterization, and photocatalysis enhancement. *Journal of Materials Research and Technology, 23*, 2454–2466. https://doi.org/10.1016/j.jmrt.2023.05.096.
5. Amin, M. T., Alazba, A. A., & Manzoor, U. (2014). A review of removal of pollutants from water/wastewater using different types of nanomaterials. *Advances in Materials Science and Engineering, 2014*, 1–24. https://doi.org/10.1155/2014/825910.
6. Ao, B., Qiu, R., & Hu, S. X. (2018). First-principles insights into the oxidation states and electronic structures of ceria-based binary, ternary, and quaternary oxides. *The Journal of Physical Chemistry C, 123*(1), 175–184. https://doi.org/10.1021/acs.jpcc.8b08622.
7. Ashwin Kishore, M. R., & Ravindran, P. (2017). Tailoring the electronic band gap and band edge positions in the C_2N monolayer by P and as substitution for photocatalytic water splitting. *The Journal of Physical Chemistry C, 121*(40), 22216–22224. https://doi.org/10.1021/acs.jpcc.7b07517.
8. Borges, M. E., Sierra, M., Cuevas, E., García, R. D., & Esparza, P. (2016). Photocatalysis with solar energy: Sunlight-responsive photocatalyst based on TiO_2 loaded on a natural material for wastewater treatment. *Solar*

Energy, 135, 527–535. https://doi.org/10.1016/j.solener.2016.06.032.

9. Camacho-Muñoz, D., Lawton, L. A., & Edwards, C. (2020). Degradation of okadaic acid in seawater by UV/TiO_2 photocatalysis–Proof of concept. *Science of the Total Environment, 733*, 139346. https://doi.org/10.1016/j.scitotenv.2020.139346.
10. Chen, H., & Wang, L. (2014). Nanostructure sensitization of transition metal oxides for visible-light photocatalysis. *Beilstein Journal of Nanotechnology, 5*(1), 696–710. https://doi.org/10.3762/bjnano.5.74.
11. Chen, P., Liu, H., Cui, W., Lee, S. C., Wang, L. A., & Dong, F. (2020). Bi-based photocatalysts for light-driven environmental and energy applications: Structural tuning, reaction mechanisms, and challenges. *EcoMat, 2*(3), 12047. https://doi.org/10.1002/eom2.12047.
12. Chu, L., Wang, J., He, S., Chen, C., Wojnárovits, L., & Takács, E. (2021). Treatment of pharmaceutical wastewater by ionizing radiation: Removal of antibiotics, antimicrobial resistance genes and antimicrobial activity. *Journal of Hazardous Materials, 415*, 125724. https://doi.org/10.1016/j.jhazmat.2021.125724.
13. Dake, D. V., Raskar, N. D., Mane, V. A., Sonpir, R. B., Stathatos, E., Asokan, K., & Dole, B. N. (2020). Exploring the role of defects on diverse properties of Cr-substituted ZnS nanostructures for photocatalytic applications. *Applied Physics A, 126*, 1–15. https://doi.org/10.1007/s00339-020-3475-0.
14. Dlamini, L. N., Krause, R. W., Kulkarni, G. U., & Durbach, S. H. (2011). Synthesis and characterization of titania-based binary metal oxide nanocomposite as potential environmental photocatalysts. *Materials Chemistry and Physics, 129*(1–2), 406–410. https://doi.org/10.1016/j.matchemphys.2011.05.029.
15. Edalati, P., Wang, Q., Razavi-Khosroshahi, H., Fuji, M., Ishihara, T., & Edalati, K. (2020). Photocatalytic hydrogen evolution on a high-entropy oxide. *Journal of Materials Chemistry A, 8*(7), 3814–3821. https://doi.org/10.1039/C9TA11479F.
16. Fenoll, J., Flores, P., Hellín, P., Martínez, C. M., & Navarro, S. (2012). Photodegradation of eight miscellaneous pesticides in drinking water after treatment with semiconductor materials under sunlight at pilot plant scale. *Chemical Engineering Journal, 204*, 54–64. https://doi.org/10.1016/j.cej.2012.06.049.
17. Fox, M. A. (1983). Organic heterogeneous photocatalysis: Chemical conversions sensitized by irradiated semiconductors. *Accounts of Chemical Research, 16*(9), 314–321. https://doi.org/10.1021/ar00083a001.

18. Gonuguntla, S., Kamesh, R., Pal, U., & Chatterjee, D. (2023). Dye sensitization of TiO_2 relevant to photocatalytic hydrogen generation: Current research trends and prospects. *Journal of Photochemistry and Photobiology C: Photochemistry Reviews, 4*(3), 100621. https://doi.org/10.1016/j.jphotochemrev.2023.100621.
19. Guo, Q., Zhou, C., Ma, Z., & Yang, X. (2019). Fundamentals of TiO_2 photocatalysis: Concepts, mechanisms, and challenges. *Advanced Materials, 31*(50), 1901997. https://doi.org/10.1002/adma.201901997.
20. Han, C., Yang, M. Q., Weng, B., & Xu, Y. J. (2014). Improving the photocatalytic activity and anti-photocorrosion of semiconductor ZnO by coupling with versatile carbon. *Physical Chemistry Chemical Physics, 16*(32), 16891–16903. https://doi.org/10.1039/C4CP01242H.
21. He, X., Wang, A., Wu, P., Tang, S., Zhang, Y., Li, L., & Ding, P. (2020). Photocatalytic degradation of microcystin-LR by modified TiO_2 photocatalysis: A review. *Science of the Total Environment, 743*, 140694. https://doi.org/10.1016/j.scitotenv.2020.140694.
22. Hu, S. W., Yang, L. W., Tian, Y., Wei, X. L., Ding, J. W., Zhong, J. X., & Chu, P. K. (2014). Non-covalent doping of graphitic carbon nitride with ultrathin graphene oxide and molybdenum disulfide nanosheets: An effective binary heterojunction photocatalyst under visible light irradiation. *Journal of Colloid and Interface Science, 431*, 42–49. https://doi.org/10.1016/j.jcis.2014.08.037.
23. Huang, F., Yan, A., & Zhao, H. (2016). Influences of doping on photocatalytic properties of TiO_2 photocatalyst. In *Semiconductor Photocatalysis—Materials, Mechanisms, and Applications* (Vol. 4, pp. 31–80). https://doi.org/10.1016/B978-0-12-803587-5.00003-5.
24. Jaiswal, R., Patel, N., Dashora, A., Fernandes, R., Yadav, M., Edla, R., & Miotello, A. (2016). Efficient Co-B-codoped TiO_2 photocatalyst for degradation of organic water pollutant under visible light. *Applied Catalysis B: Environmental, 183*, 242–253. https://doi.org/10.1016/j.apcatb.2015.10.012.
25. Jiang, L., Yuan, X., Pan, Y., Liang, J., Zeng, G., Wu, Z., & Wang, H. (2017). Doping of graphitic carbon nitride for photocatalysis: A review. *Applied Catalysis B: Environmental, 217*, 388–406. https://doi.org/10.1016/j.apcatb.2017.06.022.
26. Kambur, A., Pozan, G. S., & Boz, I. (2012). Preparation, characterization, and photocatalytic activity of TiO_2–ZrO_2 binary oxide nanoparticles. *Applied Catalysis B: Environmental, 115*, 149–158. https://doi.org/10.1016/j.apcatb.2011.12.018.

27. Keller, V., & Garin, F. (2003). Photocatalytic behavior of a new composite ternary system: WO_3/SiC-TiO_2. Effect of the coupling of semiconductors and oxides in photocatalytic oxidation of methylethylketone in the gas phase. *Catalysis Communications, 4*(8), 377–383. https://doi.org/10.1016/S1566-7367(03)00067-7.

28. Khairy, M., & Zakaria, W. (2014). Effect of metal-doping of TiO_2 nanoparticles on their photocatalytic activities toward removal of organic dyes. *Egyptian Journal of Petroleum, 23*(4), 419–426. https://doi.org/10.1016/j.ejpe.2014.01.002.

29. Krishnan, A., Swarnalal, A., Das, D., Krishnan, M., Saji, V. S., & Shibli, S. M. A. (2024). A review on transition metal oxides based photocatalysts for degradation of synthetic organic pollutants. *Journal of Environmental Sciences, 139*, 389–417. https://doi.org/10.1016/j.jes.2023.10.002.

30. Kudo, A., & Sekizawa, M. (2000). Photocatalytic H_2 evolution under visible light irradiation on Ni-doped ZnS photocatalyst. *Chemical Communications,* (15), 1371–1372. https://doi.org/10.1039/B002380F.

31. Lan, W. U., Gang, G. E., & Jinbao, W. A. N. (2009). Biodegradation of oil wastewater by free and immobilized Yarrowia lipolytica W29. *Journal of Environmental Sciences, 21*(2), 237–242. https://doi.org/10.1016/S1001-0742(08)62304-6.

32. Li, X., Zhao, W., & Zhao, J. (2002). Visible light-sensitized semiconductor photocatalytic degradation of 2,4-dichlorophenol. *Science in China Series B: Chemistry, 45*, 421–425. https://doi.org/10.1360/02yb9036.

33. Li, Y., He, R., Han, P., Hou, B., Peng, S., & Ouyang, C. (2020). A new concept: Volume photocatalysis for efficient H_2 generation using low polymeric carbon nitride as an example. *Applied Catalysis B: Environmental, 279*, 119379. https://doi.org/10.1016/j.apcatb.2020.119379.

34. Liu, N., Chen, X., Zhang, J., & Schwank, J. W. (2014). A review on TiO_2-based nanotubes synthesized via hydrothermal method: Formation mechanism, structure modification, and photocatalytic applications. *Catalysis Today, 225*, 34–51. https://doi.org/10.1016/j.cattod.2013.08.016.

35. Magdalane, C. M., Kaviyarasu, K., Vijaya, J. J., Siddhardha, B., & Jeyaraj, B. (2016). Photocatalytic activity of binary metal oxide nanocomposites of CeO_2/CdO nanospheres: Investigation of optical and antimicrobial activity. *Journal of Photochemistry and Photobiology B: Biology, 163*, 77–86. https://doi.org/10.1016/j.jphotobiol.2016.06.001.

36. Mandade, P. (2021). Introduction, basic principles, mechanism, and challenges of photocatalysis. In *Handbook of Nanomaterials for Wastewater Treatment* (Vol. 3, pp. 137–154). Elsevier. https://doi.

org/10.1016/B978-0-12-818705-5.00008-2.

37. Marcolongo, D. M. S., Nocito, F., Ditaranto, N., Comparelli, R., Aresta, M., & Dibenedetto, A. (2022). Opto-electronic characterization of photocatalysts based on p-n junction ternary and quaternary mixed oxides semiconductors (Cu_2O-In_2O_3 and Cu_2O-In_2O_3-TiO_2). *Catalysts, 12*(2), 153. https://doi.org/10.3390/catal12020153.

38. Melvin, S. D., & Leusch, F. D. (2016). Removal of trace organic contaminants from domestic wastewater: A meta-analysis comparison of sewage treatment technologies. *Environment International, 92*, 183–188. https://doi.org/10.1016/j.envint.2016.03.010.

39. Meng, X., & Zhang, Z. (2016). Bismuth-based photocatalytic semiconductors: Introduction, challenges, and possible approaches. *Journal of Molecular Catalysis A: Chemical, 423*, 533–549. https://doi.org/10.1016/j.molcata.2016.02.019.

40. Monfort, O., & Petrisková, P. (2021). Binary and ternary vanadium oxides: General overview, physical properties, and photochemical processes for environmental applications. *Processes, 9*(2), 214. https://doi.org/10.3390/pr9020214.

41. Mustafa, Y. A., Jaid, G. M., Alwared, A. I., & Ebrahim, M. (2014). The use of artificial neural network (ANN) for the prediction and simulation of oil degradation in wastewater by AOP. *Environmental Science and Pollution Research, 21*, 7530–7537. https://doi.org/10.1007/s11356-014-2693-3.

42. Neppolian, B., Wang, Q., Yamashita, H., & Choi, H. (2007). Synthesis and characterization of ZrO_2–TiO_2 binary oxide semiconductor nanoparticles: Application and interparticle electron transfer process. *Applied Catalysis A: General, 333*(2), 264–271. https://doi.org/10.1016/j.apcata.2007.01.021.

43. Noureldine, D., & Takanabe, K. (2016). State-of-the-art Sn^{2}□-based ternary oxides as photocatalysts for water splitting: Electronic structures and optoelectronic properties. *Catalysis Science & Technology, 6*(21), 7656–7670. https://doi.org/10.1039/C6CY01697E.

44. Nur, A. S., Sultana, M., Mondal, A., Islam, S., Robel, F. N., Islam, A., & Sumi, M. S. A. (2022). A review on the development of elemental and codoped TiO_2 photocatalysts for enhanced dye degradation under UV–vis irradiation. *Journal of Water Process Engineering, 47*, 102728. https://doi.org/10.1016/j.jwpe.2022.102728.

45. Ohno, T., Miyamoto, Z., Nishijima, K., Kanemitsu, H., & Xueyuan, F. (2006). Sensitization of photocatalytic activity of S- or N-doped TiO_2 particles by adsorbing Fe^{3}□ cations. *Applied Catalysis A: General, 302*(1), 62–68. https://doi.org/10.1016/j.apcata.2006.01.013.

46. Ohtani, B. (2010). Photocatalysis A to Z—What we know and what we do not know in a scientific sense. *Journal of Photochemistry and Photobiology C: Photochemistry Reviews, 11*(4), 157–178. https://doi.org/10.1016/j.jphotochemrev.2010.08.001.

47. Ohtani, B. (2011). Photocatalysis by inorganic solid materials: Revisiting its definition, concepts, and experimental procedures. In *Advances in Inorganic Chemistry* (Vol. 63, pp. 395–430). Academic Press. https://doi.org/10.1016/B978-0-12-387674-4.00008-7.

48. Panda, B., Mondal, D., Mandal, S., Khatun, J., Mukherjee, A., & Dhak, D. (2023). One-pot solution combustion synthesis of porous spherical-shaped magnesium zinc binary oxide for efficient fluoride removal and photocatalytic degradation of methylene blue and Congo red dye. *Environmental Science and Pollution Research, 30*(34), 81386–81402. https://doi.org/10.1007/s11356-023-28505-3.

49. Petit, M., Michez, L., Raimundo, J. M., Malinowski, T., & Dumas, P. (2016). An introduction to photocatalysis through methylene blue photodegradation. *European Journal of Physics, 37*(6), 065808. https://doi.org/10.1088/0143-0807/37/6/065808.

50. Qasem, N. A., Mohammed, R. H., & Lawal, D. U. (2021). Removal of heavy metal ions from wastewater: A comprehensive and critical review. *Npj Clean Water, 4*(1), 1–15. https://doi.org/10.1038/s41545-021-00127-0.

51. Rajeshwar, K., Hossain, M. K., Macaluso, R. T., Janáky, C., Varga, A., & Kulesza, P. J. (2018). Copper oxide-based ternary and quaternary oxides: Where solid-state chemistry meets photo electrochemistry. *Journal of The Electrochemical Society, 165*(4), H3192–H3201. https://doi.org/10.1149/2.0221804jes.

52. Rashed, M. N. (2013). Adsorption technique for the removal of organic pollutants from water and wastewater. In M. N. Rashed (Ed.), *Organic Pollutants-Monitoring, Risk and Treatment* (pp. 167–194). IntechOpen. https://doi.org/10.5772/54048.

53. Ravelli, D., Dondi, D., Fagnoni, M., & Albini, A. (2009). Photocatalysis. A multi-faceted concept for green chemistry. *Chemical Society Reviews, 38*(7), 1999–2011. https://doi.org/10.1039/B714486B.

54. Reginato, G., Zani, L., Calamante, M., Mordini, A., & Dessì, A. (2020). Dye-sensitized heterogeneous photocatalysts for green redox reactions. *European Journal of Inorganic Chemistry, 2020*(11–12), 899–917. https://doi.org/10.1002/ejic.202000172.

55. Sheng, W., Huang, F., & Lang, X. (2023). NH_2-MIL-125(Ti)/amorphous

TiO_2 microspheres for enhanced visible light photocatalytic selective oxidation of amines. *Materials Today Chemistry, 30*, 101505. https://doi.org/10.1016/j.mtchem.2023.101505.

56. Siao, C. W., Lee, W. L. W., Dai, Y. M., Chung, W. H., Hung, J. T., Huang, P. H., & Chen, C. C. (2019). BiOxCly/BiOmBrn/BiOpIq/GO quaternary composites: Syntheses and application of visible-light-driven photocatalytic activities. *Journal of Colloid and Interface Science, 544*, 25–36. https://doi.org/10.1016/j.jcis.2019.03.078.
57. Sinar Mashuri, S. I., Ibrahim, M. L., Kasim, M. F., Mastuli, M. S., Rashid, U., Abdullah, A. H., & Yun Hin, T. Y. (2020). Photocatalysis for organic wastewater treatment: From the basis to current challenges for society. *Catalysts, 10*(11), 1260. https://doi.org/10.3390/catal10111260.
58. Sobahi, T. R., & Amin, M. S. (2019). Upgrading the photocatalytic achievement of gC_3N_4 nanosheets along decoration with Ag@TiO_2 nanospheres for the preparation of vitamin B3. *Applied Nanoscience, 9*, 1621–1636. https://doi.org/10.1007/s13204-018-0772-2.
59. Sotto, A., Kim, J., Arsuaga, J. M., Del Rosario, G., Martínez, A., Nam, D., & Van der Bruggen, B. (2014). Binary metal oxides for composite ultrafiltration membranes. *Journal of Materials Chemistry A, 2*(19), 7054–7064. https://doi.org/10.1039/C4TA00203G.
60. Suhaimi, N. A. A., Kong, C. P. Y., Shahri, N. N. M., Nur, M., Hobley, J., & Usman, A. (2022). Dynamics of diffusion- and immobilization-limited photocatalytic degradation of dyes by metal oxide nanoparticles in binary or ternary solutions. *Catalysts, 12*(10), 1254. https://doi.org/10.3390/catal12101254.
61. Suresh, R., Rajendran, S., Kumar, P. S., Vo, D. V. N., & Cornejo-Ponce, L. (2021). Recent advancements of spinel ferrite based binary nanocomposite photocatalysts in wastewater treatment. *Chemosphere, 274*, 129734. https://doi.org/10.1016/j.chemosphere.2021.129734.
62. Tahir, M., Tasleem, S., & Tahir, B. (2020). Recent development in band engineering of binary semiconductor materials for solar driven photocatalytic hydrogen production. *International Journal of Hydrogen Energy, 45*(32), 15985–16038. https://doi.org/10.1016/j.ijhydene.2020.04.178.
63. Toumazatou, A., Arfanis, M. K., Pantazopoulos, P. A., Kontos, A. G., Falaras, P., Stefanou, N., & Likodimos, V. (2017). Slow-photon enhancement of dye sensitized TiO_2 photocatalysis. *Materials Letters, 197*, 123–126. https://doi.org/10.1016/j.matlet.2017.03.034.
64. Vaz, B., & Pérez-Lorenzo, M. (2023). Unraveling structure-performance

relationships in porphyrin-sensitized TiO_2 photocatalysts. *Nanomaterials, 13*(6), 1097. https://doi.org/10.3390/nano13061097.

65. Wang, C. Y., Pagel, R., Dohrmann, J. K., & Bahnemann, D. W. (2006). Antenna mechanism and deaggregation concept: Novel mechanistic principles for photocatalysis. *Comptes Rendus Chimie, 9*(5–6), 761–773. https://doi.org/10.1016/j.crci.2005.12.012.
66. Wang, J., Hasegawa, T., Asakura, Y., & Yin, S. (2022). Recent advances in ternary metal oxides modified by N atom for photocatalysis. *Catalysts, 12*(12), 1568. https://doi.org/10.3390/catal12121568.
67. Xiao, G., Wang, X., Li, D., & Fu, X. (2008). $InVO_4$-sensitized TiO_2 photocatalysts for efficient air purification with visible light. *Journal of Photochemistry and Photobiology A: Chemistry, 193*(2–3), 213–221. https://doi.org/10.1016/j.jphotochem.2007.07.001.
68. Xu, J. H., Li, J., Dai, W. L., Cao, Y., Li, H., & Fan, K. (2008). Simple fabrication of twist-like helix N, S-codoped titania photocatalyst with visible-light response. *Applied Catalysis B: Environmental, 79*(1), 72–80. https://doi.org/10.1016/j.apcatb.2007.09.040.
69. Yuan, Y., Lv, J., Jiang, X., Li, Z., Yu, T., Zou, Z., & Ye, J. (2007). Large impact of strontium substitution on photocatalytic water splitting activity of $BaSnO_3$. *Applied Physics Letters, 91*(9), 087404. https://doi.org/10.1063/1.2767224.
70. Zhang, J., Ma, Y., Du, Y., Jiang, H., Zhou, D., & Dong, S. (2017). Carbon nanodots/WO_3 nanorods Z-scheme composites: Remarkably enhanced photocatalytic performance under broad spectrum. *Applied Catalysis B: Environmental, 209*, 253–264. https://doi.org/10.1016/j.apcatb.2017.03.007.
71. Zhao, Z., Cong, B., Su, Z., & Li, B. (2020). Self-assembly of biarsenate capped Keggin arsenomolybdates with tetravanadium substitution for photocatalytic degradation of organic dyes. *Crystal Growth & Design, 20*(4), 2753–2760. https://doi.org/10.1021/acs.cgd.9b01701.

CHAPTER 6

PHOTO-ACTIVATED PERSULFATE-BASED AOPS

LEARNING OBJECTIVES

At the end of this chapter, readers will be able to:

1. Understand the fundamentals of photo-activated persulfate-based Advanced Oxidation Processes (AOPs) for wastewater treatment.
2. Comprehend how persulfate radicals are generated and utilized to degrade organic pollutants in photo-activated persulfate-based AOPs.
3. Identify the different types of photo-activated persulfate-based AOPs, including UV-Based, Fe-based, and TiO2-based systems, and their respective mechanisms.
4. Understand the future prospects of photo-activated PS-AOPs.

INTRODUCTORY EXAMPLE

AquaClean Solutions recently collaborated with a municipal wastewater treatment plant struggling to meet effluent quality standards due to persistent organic pollutants and emerging contaminants. Conventional treatment methods were proving insufficient, prompting the exploration of advanced oxidation processes (AOPs) for enhanced pollutant removal.

To tackle this challenge, AquaClean Solutions proposed the implementation of photo-activated persulfate-based AOPs, leveraging the powerful oxidative capabilities of persulfate radicals to degrade organic pollutants effectively. By harnessing the synergy between persulfate activation and ultraviolet (UV) radiation or transition metal catalysts, AquaClean Solutions aimed to elevate treatment performance and achieve regulatory compliance.

Through a pilot-scale demonstration at the wastewater treatment plant, AquaClean Solutions showcased the efficacy of photo-activated persulfate-based AOPs in significantly reducing pollutant concentrations and improving effluent quality. The successful implementation not only underscored the potential of AOPs in wastewater treatment but also demonstrated AquaClean Solutions' commitment to sustainable water management practices.

As AquaClean Solutions continues to innovate and optimize photo-activated persulfate-based AOPs, the company remains dedicated to providing cost-effective, environmentally friendly solutions that safeguard water resources and promote community well-being.

UNIT INTRODUCTION

In the present era, due to the swift advancement of industry, there has been a surge in the presence of environmental pollutants in water. This situation poses a global threat to both humans and aquatic organisms. The swift advancement of the dyeing and printing sector has led to the progressive substitution of raw materials with chemical dyed materials, thereby significantly augmenting the complexity of treating print and dye wastewater. Pharmaceuticals and personal care products, also known as PPCPs, are a growing concern as they have been detected in aquatic environments around the world at varying concentrations (Yang et al., 2021). The long-term presence of PPCPs in these environments can lead to physiologic toxicity in aquatic animals and contribute to the development of antibiotic resistance in bacteria. Endocrine Disrupting Chemicals (EDCs) in water ecosystems are a collection of developing contaminants that pose harm to the physiological endocrine system. These compounds are obtained from many sources such as plasticizers, pesticides, medicines, flame retardants, and similar substances. Many harmful substances are persistent organic pollutants (POPs) that can survive in aquatic settings for extended periods, increasing their chemical toxicity. Prior research has demonstrated that the molecular compositions of numerous persistent organic pollutants (POPs) exhibit a high level of stability, rendering them resistant to degradation by conventional biological treatment methods. Consequently, there is an urgent need for effective and economical water treatment technology to safeguard valuable aquatic ecosystems (Ge et al., 2021).

It is important to highlight that among the ways used to activate PS, renewable and sustainable photo-activation can effectively remove different types of organic contaminants in water. This approach works well under mild settings and is easy to operate. Photo-activated PS-AOPs have gained significant attention in recent years as a viable method for efficiently breaking down organic pollutants in water. This strategy is particularly attractive due to its cost-effectiveness and simplicity of use. Historically, researchers mostly utilized direct UV irradiation to activate PS through photo-activation, as it possesses a high energy level (Sözüdoğru et al., 2023). Nevertheless, ultraviolet (UV) light comprises merely 3% of the total sunshine.

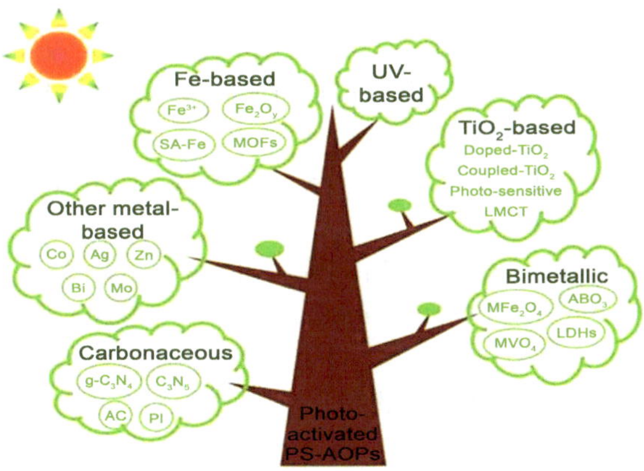

Figure 6.1. *Illustration of the overall strategies of photo-activated PS-AOPs.*

Source: Chun Zao, Creative Commons License.

Utilizing sunlight, visible light, and near IR radiation in photo-activated PS-AOPs can enhance sustainability for both the economy and the environment. Currently, the focus of research is on incorporating UV/Vis/NIR-responsive catalysts into photocatalyst-based photo-activated PS-AOPs in order to enhance pollutant elimination and mineralization efficiency. Photocatalytic events take place in photoactivated PS-AOPs, where charge separation is triggered in photo-responsive semiconductors through excitation with ultra-bandgap light. With the exception of direct photo-activation, the utilization of separated charge carriers may be employed to stimulate PS and produce reactive oxygen species (ROS) with a significant oxidation potential (Li et al., 2022) (Figure 6.1).

The purpose of this chapter is to provide a concise overview of the accomplishments, difficulties, and potential of catalyst-based photo-activated PS-AOPs. Additionally, it aims to inspire the development of innovative and potential PS-AOPs for water treatment (Tian et al., 2022).

Did You Know?

Photo-activated persulfate-based AOPs offer a chemical-free and environmentally friendly approach to wastewater treatment.

6.1. FUNDAMENTALS OF PHOTO-ACTIVATED PS-AOPS

Learning Objectives:

- Understand the mechanism of photo-activated persulfate-based AOPs
- Comprehend the chemical reactions involved in the activation of persulfate molecules.

PS can readily dissolve in water in practical scenarios. At a temperature of 20°C, the dissolution of PDS in water is approximately 520 grams per liter, while the solubility of PMS is over 250 grams per liter. PDS and PMS each have a molecular mass of 192.0 grams and 113.0 grams, respectively. PDS has an even distribution of charge and a symmetrical shape, but PMS is not symmetrical and has a slightly positive charge on the oxygen bonded to the hydrogen. This leads to a disparity in the distribution of charges between PDS and PMS. Furthermore, the bonding energy of the O–O bonding in PDS and PMS is 140.0.0 and 377 kilo Joules per mole, respectively. This suggests that activation mechanisms involving the transfer of energy, such as heating and ultraviolet radiation, are more effective in PDS (He et al., 2022). This is because PDS has a lower O–O bonding energy compared to PMS. In general, the primary objective of PS-AOPs is to liberate a significant quantity of SO_4^- by breaking the peroxide bond (O–O) of PDS or PMS through either homolytic or heterolytic cleavage. One way to break the O-O bond is by injecting energy by either UV photolysis or thermolysis, which involve the transmission of photons or heat, respectively. Alternatively, electron transfer processes can cause the breakage of the O-O link in a PS molecule when using reduced metals, oxides of metal, and their combined forms. The energy generated by heat or UV irradiation can particularly break the O–O bond of PMS and PDS and yield SO_4^-. In photo-activated PS activation systems that rely on catalysts, when the semiconductor becomes excited, electrons that are in the valence band move to the conduction band, resulting in the creation of holes in the VB. The cavities have the ability to cause water that has been absorbed on the photocatalyst's surface to undergo oxidation, resulting in the formation of hydroxyl radicals. Additionally, the electrons have the potential to engage with the absorbed oxygen, resulting in the creation of a superoxide radical (O_2^-). When PS is present, the electrons that are stimulated by light selectively move to activate the O-O bond in PS because of their greater tendency to undergo chemical reactions. This results in the creation of SO_4^-. In addition, the outermost layer of catalysts can induce changes in the valence of metal, resulting in the generation of free electrons that can activate PS. Therefore, in photo-activated PS-AOPs, the production of HO, O_2^- and SO_4^- may all lead to the breakdown of organic pollutants (Liu et al., 2023).

$$S_2O_8^{2-} \rightarrow 2SO_4^{\bullet -}$$

$$HSO_5^- \rightarrow SO_4^{\bullet -} + HO^{\bullet}$$

$$\text{Photocatalyst} + h\nu \rightarrow \text{Photocatalyst}\ (e^- + h^+$$

$$h^+ + H_2O \rightarrow HO^{\bullet} + H^+$$

$$e^- + O_2 \rightarrow O_2^{\bullet -}$$

$$HSO_5^- + e^- \rightarrow SO_4^{\bullet -} + OH^-$$

$$HSO_5^- + e^- \rightarrow SO_4^{2-} + HO^{\bullet}$$

$$S_2O_8^{2-} + e^- \rightarrow SO_4^{\bullet -} + SO_4^{2-}$$

$$S_2O_8^{2-} + M^n \rightarrow M^{n+1} + SO_4^{2-} + SO_4^{\bullet -}$$

$$HSO_5^- + M^n \rightarrow M^{n+1} + SO_4^{\bullet -} + OH^-$$

The primary mechanisms of photo-activated PS activation in the published investigations include dye-photosensitization activation, ligand-to-metal charge transfer (LMCT) activation, as well as photocatalytic activation. Due to the fact that UV radiation constitutes just 3% of the whole sunlight spectrum, a method called dye photosensitization activation has been rapidly developed. This method involves using photo-sensitive organic pigments that respond to visible light in order to activate the photosensitizer (PS). More precisely, when visible light is directed at the dye molecules, electrons from the HOMO can be stimulated and move to the LUMO. These electrons are then captured by PS, as shown in Figure 6.2(a). In the process of dye-photosensitization activation, the photosensitive dyes play a dual role as both the photosensitizer and the material to be broken down (Chen et al., 2024).

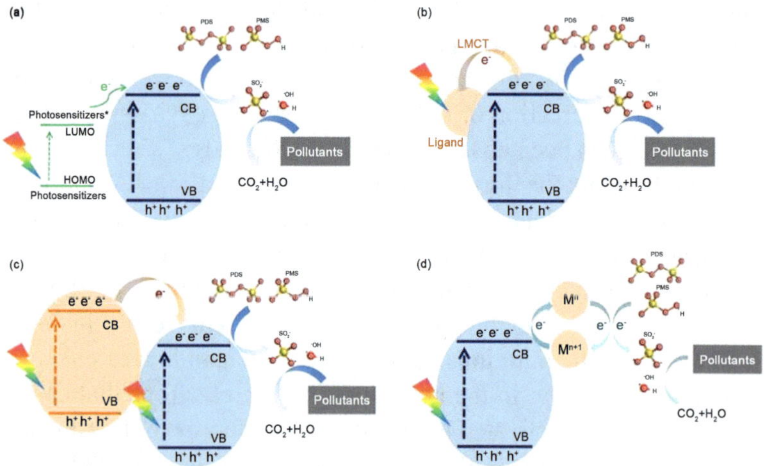

Figure 6.2. *Schematic of reaction pathways of (a) dye-sensitized; (b) LMCT process; (c) heterojunction photocatalysis; and (d) in-situ metal photo-reduction in photo-activated PS-AOPs.*

Source: Mingshan Zhu, Creative Commons License.

The LMCT procedure is a frequently used photosensitization technique to activate PS in photo-activated PS-AOPs. Specific organic pollutants can function as ligands, forming surface charge-transfer compounds with broad band-gap semiconductors. This interaction allows these semiconductors to be activated by visible light. When exposed to visible light, electrons from specific organic pollutants enter the conduction band of photocatalysts by the process of ligand-to-metal charge transfer (LMCT) (Eskandarian et al., 2021). Subsequently, the PS captures the electrons to produce reactive oxygen species (ROS) in order to break down the pollutants (Figure 6.2(b)). Within photocatalytic activation, the electrons created by light can move to the conduction band (CB) of the photocatalyst and then react with the O–O bond of the photosensitizer (PS) to generate $SO_4^{\cdot-}$. Due to the limited lifetimes of photo-generated electron-hole pairs, which impede the use of photocatalysts, there has been extensive research on the development of heterojunction photocatalysts. In these systems, photogenerated pairs of electrons and holes are initially produced in both photo-excited semiconductors. Afterward, the electrons from a semiconductor with a greater conduction band (CB) energy can move to a semiconductor having a lower CB energy. This process helps to separate hole-electron pairs and increase the activation of the photosystem (PS), as shown in Figure 6.2(c). In addition, the combination of transition metal species with photocatalysts to create in-situ metal photo reduction is advantageous for facilitating the valence cycle of transition metals. This, in turn, enables the recycling and efficient exploitation of the transition metals in the activation of PS. More precisely, following the absorption of light, a portion of the electrons that are excited by the photons in the photocatalysts are transferred to the transition metals species. This transfer facilitates the regeneration of the reductive transition metal, enabling the ongoing activation of thc photoscnsitizcr (as shown in Figurc 6.2(d)) (Khajouci et al., 2022).

6.2. UV-BASED PHOTO-ACTIVATED PS-AOPS

Learning Objectives:

- Comprehend the mechanism of UV-based photo-activated persulfate-based AOPs for wastewater treatment.

UV-activated PS-AOPs have shown to be an efficient and uncomplicated method for eliminating organic pollutants in water since the implementation of PS activation technology in water treatment. Multiple comprehensive studies have indicated that utilizing renewable UV irradiation with a wavelength range of 200–400 nm, which possesses high energy, is a very effective and environmentally conscious approach for activating PS in the treatment of water (Giannakis et al., 2017). In general, the breakdown of organic pollutants using UV-activated PS systems involves two routes. UV energy input has the ability to directly destroy some pollutants by disrupting their chemical connections. Alternatively, UV light can disrupt the peroxide link in PDS or PMS, resulting in the formation of SO_4^- ions. These ions can then oxidize water producing HO, leading to the degradation of chemical bonds in different organic pollutants present in water (Wu et al., 2024).

Dionysiou's research team discovered that the PDS/UV procedure effectively decomposed 93.2% of lindane in a time frame of 180 minutes. This degradation occurred through a reaction involving the removal of a hydrogen atom by SO_4^- and the breaking of the carbon-chlorine bond under the influence of UV irradiation. Following the PDS/UV treatment, a reduced number of by-products, which were more hazardous than lindane, were generated. These by-products could be effectively broken down under strong UV exposure.

In a separate investigation, it was found that the PDS/UV system effectively eliminated oxytetracycline (OTC) after 600 minutes (Dibene et al., 2020). The major active species at a pH of 7 was SO_4^-, while the concentration of HO steadily increased from pH 9 to 11.35. Furthermore, the PDS/UV treatment resulted in the degradation of 91.0% of sulfamethoxypyridazine (SMP) after 10 minutes, primarily due to the generation of SO_4^- and HO. The rate of degradation of SMP was observed to be strongly influenced by pH, with a decreasing trend observed in the following order: pH 5.5 > 7.0 > 9.0. The PMS/UV system employed direct UV photolysis and reactive radicals (SO_4^- and HO) to rapidly degrade flumequine (FLU) within 60 minutes, achieving complete degradation (100%). This degradation occurred by demethylation, hydroxylation, decarboxylation, and ring-opening routes (Zhou et al., 2020). A separate investigation examined the radical chemistry involved in the breakdown of diethyl phthalate (DEP) by the PMS/UV system. The study revealed that SO_4^- exhibited more resistance to the detrimental effects of alkaline conditions, bicarbonate, and NOM compared to HO. However, SO_4^- was shown to be more susceptible to the influence of chloride than HO. UV-activated PS-AOPs offer a solution to the issues of intricate catalyst preparation and the leaking of metallic catalysts, which are present in catalyst-based PS-AOPs. Nevertheless, ultraviolet (UV) light constitutes a mere 3% of the total sunshine. Extending the use of UV rays to visible light and even NIR is crucial for achieving sustainable growth of PS-AOPs. Therefore, the future development of photo-activated PS-AOPs involves the utilization of Vis/NIR-responsive catalysts to design photocatalyst-based systems (Ghauch et al., 2017).

Remember

Factors influencing the efficiency of photo-activated persulfate-based AOPs include UV intensity, persulfate concentration, pH, and catalyst type.

6.3. FE-BASED PHOTO-ACTIVATED PERSULFATE AOPS

Learning Objectives:

- Comprehend the mechanism of Fe-based photo-activated persulfate-based AOPs for wastewater treatment.

Iron is a crucial element in chemical procedures for eliminating organic pollutants because of its abundance, eco-friendliness, and exceptional catalytic characteristics, among other transition metals (Zhou et al., 2017). Recently, several photo-activated Fe-based PS-AOPs have been created to eliminate organic contaminants in water. Ferrous iron, often in the form of dissolved Fe^{2+}, is frequently employed as a uniform transition metal for the chemically initiated activation of Fenton and PS. Nevertheless, the primary disadvantage of dissolved Fe^{2+} is its swift conversion to Fe^{3+}, resulting in the formation of ferrous hydroxide and ferric oxide sludge (Yin et al., 2020). This, in turn, obstructs Fenton and Fenton-like processes. A study found that a uniform Fe^{3+}/PMS/Vis system can facilitate inner electron transport in the HSO_5^-–$Fe(OH)^{2+}$ complex through the process of ligand-to-metal charge transfer (LMCT). The generated electrons can stimulate the production of PMS, which in turn generates HO^- and SO_4^- ions for the breakdown of 4-chlorophenol (4-CP). Furthermore, the Fe(III)–sulfamethoxazole (SMX) complex, which has a uniform composition, can initiate the PDS reaction when exposed to visible light by means of the LMCT mechanism. The complexation of Fe and SMX resulted in a significant reduction in the energy gap (Eslami et al., 2020).

It is possible to utilize lower energy visible light to stimulate the Fe(III)-SMX complex, which aids in the conversion of Iron(III) to Iron(II) for PDS activation (*see* Figure 6.3(a)). Fe-based photo-activated PS-AOPs often utilize iron oxides and their complexes as a typical approach for making the activation systems. Furthermore, a composite material consisting of Fe_2O_3/Fe_3O_4 backed by g-C_3N_4 was synthesized to catalyze the activation of PS (photosensitizer) during visible light exposure. This process effectively eliminates bisphenol A as well as dyes by producing SO_4^- and O_2^- (Macías-Vargas et al., 2022). In these instances, the presence of Fe_2O_3 and Fe_3O_4 not only decreased the energy difference between valence and conduction bands by combining with g-C_3N_4, but also efficiently decreased the recombination of electron-hole pairs formed by g-C_3N_4 as a pathway for electron transfer. The combination of b-FeOOH and g-C_3N_4 (b-FeOOH/g-C_3N_4) showed remarkable stability and efficiency in activating PMS for the elimination of carbamazepine (CBZ) during solar light exposure (Zhang et al., 2024). This process generates highly oxidizing SO_4^- and HO, which contributes to the degradation of CBZ. Contrary to a conventional PMS activation system, the oxidation process created high-valent iron (Fe(IV) QO) species through the excited-state $FePcCl_{16}$. Figure 6.3(b) demonstrates that when g-C_3N_4 and $FePcCl_{16}$ are exposed to visible light, the electrons they produce can be collected by PMS to yield SO_4^- and HO. At the same time, the active form of Fe(IV)QO, which is formed by breaking the O-O bond also showed powerful oxidative characteristics in breaking down CBZ. Atomically distributed heterogeneous metal catalysts have recently garnered significant attention in research circles due to their totally exposed active sites consisting of atomic metal. As an example, Wang's team created a catalyst with a sandwich structure consisting of C_3N_4–Fe–rGO. In this structure, the single atom (SA)-

Fe was stabilized by placing the Fe atoms between a double-layered support made of C_3N_4–rGO (An et al., 2024). The study revealed that it had a consistent PDS activation capacity for TC elimination across a broad pH spectrum ranging from 0 to 14.49.

Combining inorganic metal centers with functional organic ligands results in the formation of metal-organic frameworks (MOFs), which offer a highly customizable porous structure with the ability to adjust pore size. Fe-based MOFs, which possess significant specific surface areas, homogeneous yet modifiable cavities, and exceptional visible light absorption, have been the subject of recent intensive research (Kaur, 2020). For instance, a PDS procedure utilizing MIL-100(Fe) activated by visible light successfully eliminated 98.9% of SMX within 180 minutes. This was achieved through the in situ photoreduction of iron (III).

The electrons produced by the photocatalytic process initially converted Fe(III) into Fe(II). Subsequently, the Fe(II) produced activated the PDS to generate HO and O_2^- in order to eliminate SMX. The introduction of visible wavelengths in the MIL-100(Fe)/PDS system resulted in a 2.1-fold increase in the degrading efficiency of SMX. Iron-based photo-activated PS-AOPs have gained significant attention due to their affordability, lack of toxicity, strong catalytic performance, eco-friendliness, and excellent reusability (Ferreira et al., 2020).

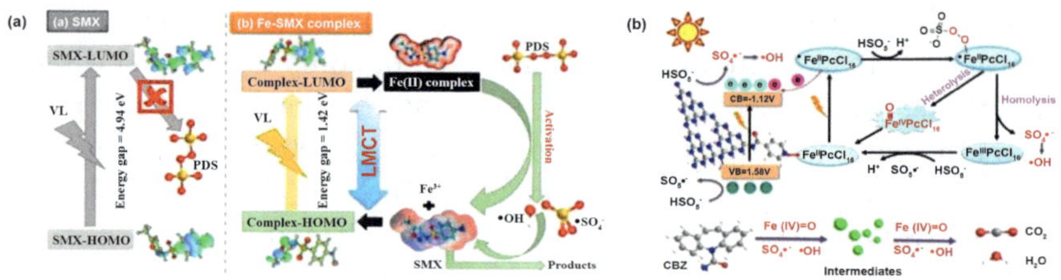

Figure 6.3. *Representation of the proposed photo-activated PS activation mechanisms.*

Source: Zhenfeng Bian, Creative Commons License.

Nevertheless, ongoing research is mostly centered around the eco-toxicology assessment of materials containing iron (Fe), whereas there remains a lack of adequate toxicology investigation into the intermediate products and by-products. Furthermore, photo-activated Fe-based PS-AOPs have the ability to generate various radicals, such as HO, SO_4^-, and O_2^-, as well as non-radicals. Investigating the assessment and control of the impact of such radicals and non-radicals is a potential field of research for the targeted breakdown of different contaminants (Guo et al., 2022).

Practice Problem

Discuss the factors influencing the efficiency of UV-Based photo-activated persulfate AOPs in degrading pollutants in wastewater. How do parameters such as UV intensity, persulfate concentration, and pH affect treatment performance?

Solution to Practice Problem

The efficiency of UV-Based photo-activated persulfate AOPs depends on various factors that influence the activation of persulfate molecules and the generation of sulfate radicals. Higher UV intensity levels generally result in more efficient photolysis of persulfate, leading to increased sulfate radical production and enhanced pollutant degradation. Similarly, higher persulfate concentrations can provide more reactive species for pollutant oxidation, but excessive concentrations may lead to scavenging reactions that reduce treatment efficiency. pH also plays a significant role, as acidic conditions promote persulfate activation and sulfate radical formation. However, extreme pH values can affect the stability of persulfate and alter the reaction kinetics. Therefore, optimizing UV intensity, persulfate concentration, and pH conditions is essential to maximize the efficiency of UV-Based photo-activated persulfate AOPs for wastewater treatment.

6.4. TIO_2-BASED PHOTO-ACTIVATED PERSULFATE AOPS

Learning Objectives:

- Understand the process of TiO2-based photo-activated persulfate-based AOPs for wastewater treatment.

Over the past few decades, TiO_2 has emerged as a favored photocatalyst for environmental cleaning due to its exceptional photoactivity, chemical inertness, affordability, and potential for efficient recycling. Nevertheless, the limited light-responsive range of TiO_2 can only be enhanced by broadening its band gap, hence enhancing its practical utility. Several techniques are considered to expand the range of TiO_2 light sensitivity to include visible light (Ahmadmoazzam et al., 2023). For example, the introduction of precious metals or non-metal components (such as nitrogen, carbon, and sulfur) into the crystalline structure of titanium dioxide has been shown to enhance the activity at low-energy bands. Furthermore, there is potential in combining TiO_2 with metallic or non-metal semiconductors in order to expand the range of light that can be detected to include the visible light region. Under visible light irradiation, electrons in the valence band (VB) of the connected semiconductor can migrate to its conduction band (CB) (Ghavi et al., 2021). These electrons then go to TiO_2's CB, where instead of acting as a photocatalyst, TiO_2 serves as an electron transfer channel.

PS can capture the electrons in the conduction band (CB) of TiO_2 to produce different reactive oxygen species (ROS) that can eliminate organic pollutants. Certain photosensitive organic pigments can be stimulated by visible light.

From the highest occupied molecular orbital (HOMO) to the lowest unoccupied molecular orbital (LUMO), the dye molecules can move photogenerated electrons. These excited electrons can then enter into the conduction band (CB) of titanium dioxide, thereby expanding the range of wavelengths that TiO_2 can absorb to include the visible area. Titanium dioxide can undergo activation when exposed to visible light through the mechanism of ligand-to-metal charge transfer (LMCT). Specific ligands, such as amino acids and antibiotic compounds, have the ability to create charge transfer complexes on the surface of different semiconductors. Several photo-activated PS-AOPs based on TiO_2 have been created to eliminate organic contaminants in water. The light-responsive properties of TiO_2 were enhanced through the implementation of metal-doping and non-metal-doping techniques. A study revealed that a Bi-TiO_2/PMS/Solar system demonstrated a higher efficiency in photocatalytic performance compared to undoped TiO_2 for the degradation of flumequine (Garg et al., 2019). This was achieved by boosting the production of hydroxyl radicals (HO) and sulfate radicals (SO_4^-). A separate work involved the attachment of N-doped TiO_2 onto acid-activated sepiolite to create a nanocomposite called N–TiO_2/ASep. This nanocomposite was used to activate PMS (persulfate) under visible light in order to degrade MO (methyl orange). The interaction between the Ti^{3+} species, leftover carbon, and intervening nitrogen inside the nanocomposite leads to an increased efficiency in separating carriers when exposed to visible light. Furthermore, sepiolite has the ability to not only initiate the production of SO_4^-, HO, and O_2^- by PMS activation but also to scatter and stabilize TiO_2 by creating Si-O-Ti linkages. Combining TiO_2 with metallic or non-metallic semiconductors can further broaden the range of light that can be detected to include the visible light spectrum (Gandhi & Prakash, 2023).

The combined actions of Fe-POMs and TiO_2 in the Fe-POMs/TiO_2 complex resulted in a BPA degradation rate that was 10 and 33 times better than that of solo photocatalysis and photoactivated PDS, correspondingly. Furthermore, the D_{35}-TiO_2/g-C_3N_4 nanocomposite expanded the range of wavelengths it can detect up to 675 nm. This is because of the combined ability of g-C_3N_4 and the D_{35} dye to absorb light. The TiO_2/PMS/Vis system is frequently employed in the decomposition of organic contaminants through the utilization of LMCT or dye-sensitization. The presence of a notable synergistic impact was seen in this system for the degradation of AO_7, achieving a 100% degradation in 100 minutes (Khasawneh & Palaniandy, 2019). This result is attributed to a process called dyesensitization. Furthermore, amino acids (AAs) can serve as visible-light sensitizers and substrates for degradation in the said system, in addition to dyes. While TiO_2 and AAs do not have the ability to absorb visible rays, the interaction between AAs and TiO_2 on the surface can result in the formation of charge-transfer complexes that are responsive to visible light. Within this study, it has been observed that AAs can undergo degradation through two distinct mechanisms: LMCT, which operates through a non-radical pathway and SO_4^-, which operates through a radical pathway. Another study has shown that PMS can act as a ligand on titanium dioxide and transfer its electrons to the conduction band (CB) of TiO_2 when exposed to visible light, using the ligand-to-metal charge transfer (LMCT) process (Xu et al., 2023).

6.5. FUTURE PROSPECTS

Learning Objectives:

- Gain insight about the suggestions for further research into photo-activated persulfate-based AOPs to expand the practical applications.

Photo-activated PS-AOPs face certain limitations that hinder their practical application. The following recommendations are provided for more research to enhance the practical implementation:

- Efforts should be made to develop synthetic technologies that are easy, cost-effective, and environmentally friendly. The majority of synthetic methods that have been documented thus far require intricate procedures or costly apparatus, which hinders the mainstream use of these approaches. Furthermore, many synthetic procedures can generate hazardous substances throughout the synthesis process. Hence, it is important to devise uncomplicated and eco-friendly methods for synthesizing catalysts (Zhi et al., 2021).
- Additional varieties of metal-free substances should be researched and created. Non-metallic materials can prevent the issue of metal leaching, as opposed to metal catalysts. Thus, it is encouraging to create metal-free materials that possess advantageous surface characteristics for photo-activated PS-AOPs. Nevertheless, the

investigation into the activation of PS without the usage of metals has primarily concentrated on g-C_3N_4 up until now.

- Photothermal catalysts can convert near-infrared light into heat, which can be used to activate photosensitizers (PS). This heat can break the O-O bond of the PS, resulting in the generation of reactive oxygen species (ROS). Therefore, it is advantageous to create photothermal catalysts that respond to near-infrared light in order to activate PS and make efficient use of energy (Gujar et al., 2023).
- The mineralization capacity of PS-AOPs, just relying on the removal of total organic carbon (TOC), is insufficient. This is because catalysts that are used in heterogeneous systems can adsorb both the parent pollutants and their intermediates, resulting in a reduction in TOC. Hence, it is important to furnish CO_2 monitoring data throughout the degradation process in order to elucidate the conversion of organic carbon in forthcoming studies.
- The selection of photo-activated PS-AOPs should be based on a toxicity assessment of degradation intermediates employing both theoretical and experimental research in order to give environmental friendliness first priority. However, the current body of research on the assessment of eco-toxicology is still insufficient. As a result, it is imperative to advance the research on water environmental toxicology with respect to PS-AOP intermediate products and by-products (Wang et al., 2023).

SUMMARY

- Due to the growing presence of dyes, PPCPs (Pharmaceuticals and Personal Care Products), and EDCs (Endocrine Disrupting Chemicals) in water, the aquatic ecosystem is encountering unprecedented difficulties. PS AOPs have gained recognition as a very promising method for water treatment due to their environmentally friendly nature, ability to operate across a wide range of pH levels, and high redox potentials of sulfate radicals.
- The chapter commences by explaining the basic principles of photo-activated persulfate-based advanced oxidation processes (AOPs), emphasizing the chemical reactions that occur during the activation process of persulfate molecules using ultraviolet (UV) rays or transition metal catalysts. The reactions produce sulfate radicals ($SO_4{\bullet}^-$), which are extremely reactive substances that can oxidize organic pollutants and microbiological contaminants in wastewater.
- The primary processes of photo-activated PS-AOPs involve dye-photosensitization activation, activation by LMCT (ligand-to-metal charge transfer), and photocatalytic activation. Significant advancements have been made in TiO_2-based, alternative metal-based, Fe-based, bimetallic, and carbonaceous photo-activated PS-AOPs in recent years.

REVIEW QUESTIONS

1. Explain the mechanism of pollutant degradation in photo-activated persulfate-based Advanced Oxidation Processes (AOPs).
2. Compare and contrast the different types of photo-activated persulfate-based AOPs, including UV-Based, Fe-based, and TiO_2-based systems.
3. Discuss the factors influencing the efficiency of photo-activated persulfate-based AOPs in degrading pollutants in wastewater.
4. Provide examples of organic pollutants commonly degraded by photo-activated persulfate-based AOPs in wastewater treatment.
5. Explore the role of catalysts, such as iron and titanium dioxide, in enhancing persulfate activation and pollutant degradation in Fe-based and TiO_2-based photo-activated persulfate AOPs.

MULTIPLE CHOICE QUESTIONS

1. **What is the primary mechanism of pollutant degradation in photo-activated persulfate-based Advanced Oxidation Processes (AOPs)?**
   a. Reduction by UV radiation
   b. Oxidation by sulfate radicals
   c. Photolysis of persulfate molecules
   d. Adsorption onto catalyst surfaces
2. **Which type of catalyst is commonly used in Fe-based photo-activated persulfate AOPs to enhance persulfate activation?**
   a. Titanium dioxide (TiO_2)
   b. Platinum (Pt)
   c. Iron (Fe)
   d. Silver (Ag)
3. **What role does ultraviolet (UV) radiation play in UV-Based photo-activated persulfate AOPs?**
   a. Generating sulfate radicals
   b. Photolyzing persulfate molecules
   c. Activating transition metal catalysts
   d. Adsorbing organic pollutants
4. **What is the purpose of using titanium dioxide (TiO_2) in TiO_2-based photo-activated persulfate AOPs?**
   a. To generate sulfate radicals
   b. To activate persulfate molecules
   c. To enhance persulfate photolysis
   d. To facilitate photocatalytic reactions
5. **Which of the following pollutants can be effectively degraded by photo-activated persulfate-based AOPs?**
   a. Inorganic salts
   b. Heavy metals
   c. Organic pollutants
   d. Suspended solids
6. **What is the primary oxidizing species responsible for pollutant degradation in photo-activated persulfate-based AOPs?**
   a. Hydroxyl radicals (•OH)
   b. Sulfate radicals ($SO_4^{\bullet -}$)

c. Singlet oxygen ($^1O_2$)

d. Ozone (O_3)

Answers to Multiple Questions

1. (b); 2. (c); 3. (b); 4. (d); 5. (c); 6. (b)

REFERENCES

1. Ahmadmoazzam, M., Akbari, H., Adibzadeh, A., Pourfadakari, S., & Akbari, H. (2023). Visible-light-driven TiO_2/Fe_2O_3/Chitosan nanocomposite with promoted photodegradation of meropenem and imipenem antibiotics by peroxymonosulfate. *Environmental Technology, 7*(3), 1–12. https://doi.org/10.1080/09593330.2023.1769802.
2. An, L., Kong, X., Jiang, M., Li, W., Lv, Q., Hou, X., & Yang, T. (2024). Photo-assisted Natural Chalcopyrite Activated Peracetic Acid for Efficient Micropollutant Degradation. *Water Research, 7*(3), 121–699. https://doi.org/10.1016/j.watres.2023.121699.
3. Chen, F., Huang, X. T., Bai, C. W., Zhang, Z. Q., Duan, P. J., Sun, Y. J., & Zhang, Y. S. (2024). Advancements in heterogeneous activation of persulfates: Exploring mechanisms, challenges in organic wastewater treatment, and innovative solutions. *Chemical Engineering Journal, 4*(3), 148789. https://doi.org/10.1016/j.cej.2023.148789.
4. Dibene, K., Yahiaoui, I., Yahia Cherif, L., Aitali, S., Amrane, A., & Aissani-Benissad, F. (2020). Paracetamol degradation by photo-activated peroxydisulfate process (UV/PDS): Kinetic study and optimization using central composite design. *Water Science and Technology, 82*(7), 1404–1415. https://doi.org/10.2166/wst.2020.379.
5. Eskandarian, M. R., Ganjkhanloo, M., Rasoulifard, M. H., & Hosseini, S. A. (2021). Energy-efficient removal of acid red 14 by UV-LED/persulfate advanced oxidation process: Pulsed irradiation, duty cycle, reaction kinetics, and energy consumption. *Journal of the Taiwan Institute of Chemical Engineers, 127*, 129–139. https://doi.org/10.1016/j.jtice.2021.06.009.
6. Eslami, A., Mehdipour, F., Lin, K. Y. A., Maleksari, H. S., Mirzaei, F., & Ghanbari, F. (2020). Sono-photo activation of percarbonate for the degradation of organic dye: The effect of water matrix and identification of by-products. *Journal of Water Process Engineering, 33*, 100998. https://doi.org/10.1016/j.jwpe.2019.100998.
7. Ferreira, L. C., Castro-Alférez, M., Nahim-Granados, S., Polo-López, M.

I., Lucas, M. S., Puma, G. L., & Fernández-Ibáñez, P. (2020). Inactivation of water pathogens with solar photo-activated persulfate oxidation. *Chemical Engineering Journal, 381*, 122275. https://doi.org/10.1016/j.cej.2019.122275.

8. Gandhi, J., & Prakash, H. (2023). Photo-disinfection processes for bacterial inactivation and underlying principles for water constituents' impact: A review. *Chemical Engineering Journal Advances, 8*(7), 100482. https://doi.org/10.1016/j.ceja.2023.100482.

9. Garg, A., Singhania, T., Singh, A., Sharma, S., Rani, S., Neogy, A., & Garg, N. (2019). Photocatalytic degradation of bisphenol-A using N, Co codoped TiO_2 catalyst under solar light. *Scientific Reports, 9*(1), 765. https://doi.org/10.1038/s41598-018-37488-9.

10. Ge, M., Hu, Z., Wei, J., He, Q., & He, Z. (2021). Recent advances in persulfate-assisted TiO_2-based photocatalysis for wastewater treatment: Performances, mechanism and perspectives. *Journal of Alloys and Compounds, 888*, 161625. https://doi.org/10.1016/j.jallcom.2021.161625.

11. Ghauch, A., Baalbaki, A., Amasha, M., El Asmar, R., & Tantawi, O. (2017). Contribution of persulfate in UV-254 nm activated systems for complete degradation of chloramphenicol antibiotic in water. *Chemical Engineering Journal, 317*, 1012–1025. https://doi.org/10.1016/j.cej.2017.02.073.

12. Ghavi, A., Bagherian, G., & Vahidian, H. R. (2021). Investigating the synergistic effect of UV/PS/TiO_2 and UV/PI/TiO_2 processes on paraquat herbicide degradation in media aqueous: Statistical optimization, kinetic study, and estimation of electrical energy consumption. *Environmental Technology, 7*(3), 875–880. https://doi.org/10.1080/09593330.2020.1839547.

13. Giannakis, S., Jovic, M., Gasilova, N., Gelabert, M. P., Schindelholz, S., Furbringer, J. M., & Pulgarin, C. (2017). Iohexol degradation in wastewater and urine by UV-based Advanced Oxidation Processes (AOPs): Process modeling and by-products identification. *Journal of Environmental Management, 195*, 174–185. https://doi.org/10.1016/j.jenvman.2017.04.003.

14. Gujar, S. K., Divyapriya, G., Gogate, P. R., & Nidheesh, P. V. (2023). Environmental applications of ultrasound activated persulfate/peroxymonosulfate oxidation process in combination with other activating agents. *Critical Reviews in Environmental Science and Technology, 53*(6), 780–802. https://doi.org/10.1080/10643389.2021.1922456.

15. Guo, R., Chen, Y., Liu, B., Han, Y., Gou, J., & Cheng, X. (2022). Catalytic degradation of lomefloxacin by photo-assisted persulfate activation on

natural hematite: Performance and mechanism. *Chinese Chemical Letters, 33*(8), 3809–3817. https://doi.org/10.1016/j.cclet.2022.04.007.

16. He, S., Chen, Y., Li, X., Zeng, L., & Zhu, M. (2022). Heterogeneous photocatalytic activation of persulfate for the removal of organic contaminants in water: A critical review. *ACS ES&T Engineering, 2*(4), 527–546. https://doi.org/10.1021/acsestengg.1c00318.
17. Kaur, B. (2020). Development of Photo-Induced Persulfate-Based Processes for Efficient Application in Water Treatment. *TalTech, 3*(1), 35–38.
18. Khajouei, G., Finklea, H. O., & Lin, L. S. (2022). UV/chlorine advanced oxidation processes for degradation of contaminants in water and wastewater: A comprehensive review. *Journal of Environmental Chemical Engineering, 10*(3), 107508. https://doi.org/10.1016/j.jece.2022.107508.
19. Khasawneh, O. F. S., & Palaniandy, P. (2019). Photocatalytic degradation of pharmaceuticals using TiO based nanocomposite catalyst-review. *Civil and Environmental Engineering Reports, 29*(3), 1–33. https://doi.org/10.2478/ceer-2019-0028.
20. Li, L., Yuan, X., Zhou, Z., Tang, R., Deng, Y., Huang, Y., & Gong, D. (2022). Research progress of photocatalytic activated persulfate removal of environmental organic pollutants by metal and nonmetal based photocatalysts. *Journal of Cleaner Production, 372*, 133420. https://doi.org/10.1016/j.jclepro.2022.133420.
21. Liu, Z., Ren, X., Duan, X., Sarmah, A. K., & Zhao, X. (2023). Remediation of environmentally persistent organic pollutants (POPs) by persulfates oxidation system (PS): A review. *Science of The Total Environment, 863*, 160818. https://doi.org/10.1016/j.scitotenv.2022.160818.
22. Macías-Vargas, J. A., Díaz-Ramírez, M. L., García-Mejía, T. A., & Ramírez-Zamora, R. M. (2022). Enhanced ciprofloxacin degradation via photo-activated persulfate using the effluent of a large wastewater treatment plant. *Topics in Catalysis, 65*(9), 1128–1138. https://doi.org/10.1007/s11244-022-01476-9.
23. Sözüdoğru, O., Koçoğlu, E. S., Yılmaz, A. E., Bakirdere, S., & Komesli, O. T. (2023). The experimental design approach to removal of endocrine disrupting compounds from domestic wastewater by electrooxidation process. *Journal of Applied Electrochemistry, 53*(10), 1971–1990. https://doi.org/10.1007/s10800-023-01878-w.
24. Tian, D., Zhou, H., Zhang, H., Zhou, P., You, J., Yao, G., & Lai, B. (2022). Heterogeneous photocatalyst-driven persulfate activation process under visible light irradiation: From basic catalyst design principles to novel

enhancement strategies. *Chemical Engineering Journal, 428*, 131166. https://doi.org/10.1016/j.cej.2021.131166.

25. Wang, D., Junker, A. L., Sillanpää, M., Jiang, Y., & Wei, Z. (2023). Photo-based advanced oxidation processes for zero pollution: Where are we now? *Engineering, 23*, 19–23. https://doi.org/10.1016/j.eng.2023.04.015.
26. Wu, Z., Gong, S., Liu, J., Shi, J., & Deng, H. (2024). Progress and problems of water treatment based on UV/persulfate oxidation process for degradation of emerging contaminants: A review. *Journal of Water Process Engineering, 58*, 104870. https://doi.org/10.1016/j.jwpe.2023.104870.
27. Xu, J., Xia, W., Sheng, G., Jiao, G., Liu, Z., Wang, Y., & Zhang, X. (2023). Progress of disinfection catalysts in advanced oxidation processes, mechanisms and synergistic antibiotic degradation. *Science of The Total Environment, 9*(3), 169580. https://doi.org/10.1016/j.scitotenv.2022.169580.
28. Yang, J., Zhu, M., & Dionysiou, D. D. (2021). What is the role of light in persulfate-based advanced oxidation for water treatment? *Water Research, 189*, 116627. https://doi.org/10.1016/j.watres.2020.116627.
29. Yin, R., Chen, Y., He, S., Li, W., Zeng, L., Guo, W., & Zhu, M. (2020). In situ photoreduction of structural Fe (III) in a metal–organic framework for peroxydisulfate activation and efficient removal of antibiotics in real wastewater. *Journal of Hazardous Materials, 388*, 121996. https://doi.org/10.1016/j.jhazmat.2020.121996.
30. Zhang, L., Liu, T., Zhu, M., & Liu, Y. (2024). A novel photo-assisted activated persulfate strategy for selective oxidation of ammonia nitrogen to dinitrogen using sodium silicate and sodium sulfite as regulators. *Chemical Engineering Journal, 479*, 147542. https://doi.org/10.1016/j.cej.2023.147542.
31. Zhi, K., Li, Z., Ma, P., Tan, Y., Zhou, Y., Zhang, W., & Zhang, J. (2021). A review of activation persulfate by iron-based catalysts for degrading wastewater. *Applied Sciences, 11*(23), 11314. https://doi.org/10.3390/app112311314.
32. Zhou, K., Wang, Z., Wang, X., Jiao, G., Li, Y., Sun, S. P., & Chen, X. D. (2020). Degradation of emerging pharmaceutical micropollutants in municipal secondary effluents by low-pressure UVC-activated HSO_5− and S_2O_{82}− AOPs. *Chemical Engineering Journal, 393*, 124712. https://doi.org/10.1016/j.cej.2020.124712.
33. Zhou, L., Ferronato, C., Chovelon, J. M., Sleiman, M., & Richard, C. (2017). Investigations of diatrizoate degradation by photo-activated persulfate. *Chemical Engineering Journal, 311*, 28–36. https://doi.org/10.1016/j.cej.2016.11.098.

CHAPTER 7

ELECTROCHEMICAL OXIDATION PROCESSES

LEARNING OBJECTIVES

At the end of this chapter, readers will be able to:

1. Understand the fundamental principles of electrochemical oxidation processes.
2. Comprehend the mechanism of photoelectrochemical processes.
3. Gain knowledge about the photo electrocatalysis process for wastewater treatment.
4. Understand the hybrid combination process of PEF and PEC.
5. Learn about the applications and prospects of electrochemical oxidation processes.
6. Comprehend the benefits and drawbacks of electrochemical oxidation processes.

INTRODUCTORY EXAMPLE

Havo Pharmaceuticals is a prominent manufacturer of pharmaceutical products, running a large-scale production facility in a densely populated urban area. The manufacturing processes at the facility produce large amounts of wastewater that contain different types of organic pollutants. If this wastewater is released into nearby water bodies without treatment, it may have adverse effects on the environment and public health.

In order to tackle the issue, Havo Pharmaceuticals is currently investigating cutting-edge wastewater treatment technologies to meet environmental regulations and reduce its impact on the environment. Through collaboration with esteemed research institutions and cutting-edge technology service providers, the company has successfully implemented a highly sophisticated electrochemical oxidation process.

Through the utilization of electrochemical oxidation, Havo Pharmaceuticals strives to effectively break down organic pollutants found in its wastewater, resulting in the discharge of cleaner effluent and a decreased environmental footprint.

UNIT INTRODUCTION

Conventional methods of water treatment reduce in comparison to electrochemical processes when it comes to effectively removing dangerous and stubborn organic micropollutants. Electrochemical methods utilize highly efficient and eco-friendly techniques to generate hydroxyl radicals, either directly or indirectly. Chemical compounds known as organic micropollutants, including persistent organic pollutants (POPs), pesticides, and herbicides, have been found in significant amounts in water bodies (Barrera-Díaz et al., 2014). These pollutants have greatly increased the toxicity levels in the aquatic environment, and traditional treatment methods have failed to effectively tackle this problem. Electrochemical methods have been instrumental in tackling these notable limitations. Utilizing powerful oxidizing agents like the hydroxyl radical with its impressive oxidation potential of 2.8 V/NHE, advanced oxidation processes (AOPs) are able to achieve remarkable results. These agents are produced within the reaction medium itself. These radicals rapidly undergo reactions with organic compounds, mainly involving dehydrogenation or hydroxylation (Darvishmotevalli et al., 2019). The previous discussion focused on the electrochemical advanced oxidation process, with a particular emphasis on electrochemistry and its application in environmental electrochemistry. Electro-Fenton was the initial technology acknowledged as an electrochemical advanced oxidation process (EAOP) due to its ability to generate and employ hydroxyl radicals for the oxidation of organic compounds. This technology relies on two processes (Moreira et al., 2017):

- The process involves the electrochemical conversion of iron(III) species to iron(II) on the cathodic surface.
- Another important step is the cathodic reduction of oxygen, which results in the formation of hydrogen peroxide.

These reactions cause hydrogen peroxide to break down catalytically into hydroxyl radicals. Table 7.1 offers a concise comparison between AOPs and EAPs (Zhou & He, 2007).

Table 7.1. *Comparison of AOPs and EAPs*

Advanced Oxidation Processes	**Electrochemical Advanced Processes**
O_3 based processes	Hydrogen recovery
Photo-Fenton, Fenton, solar photo-Fenton	Electrocoagulation
Heterogenous photocatalysis: nanofibers, nanomaterials, UVA-LED	New cell/reactors design
–	Electro-Fenton, solar electro-Fenton

Source: Chaplin, Creative Commons License.

7.1. ELECTROCHEMICAL OXIDATION PROCESSES

Learning Objectives:

- Understand the fundamental principles and mechanisms of electrochemical oxidation processes.
- Learn about the significance of electrochemical oxidation processes in the treatment of wastewater.

An oxidant, which has a stronger affinity for these tiny, charged particles, receives one or more electrons voluntarily from a reductant during the electron transfer process. The process of electron transfer between the oxidant and the reductant produces a chemical change that results in compounds with an odd number of electrons in their valence shell. Because of their unpaired electron, these substances are referred to as radicals and exhibit high levels of instability and reactivity (Wu & Zhou, 2001). One of the primary goals is to elucidate the notable efficiencies attained by EAOPs in effectively eliminating natural toxins through meticulously controlled oxidation processes throughout the treatment.

Electrochemical advanced oxidation processes (EAOPs) include the interaction between specific molecules and pre-existing oxidants on the surfaces of the anode. This interaction leads to the oxidation of harmful chemicals found in wastewater. AO rapidly oxidizes natural impurities on the anode surface and facilitates the production of significant quantities of oxidants. Oxidants could impact the surface of the cathode and greatly accelerate the oxidation process across a considerable portion of the treated waste (Canizares et al., 2009). The production and amplification of oxidants relies on a diverse range of information sources, with the choice of the anode material being very important. In wastewater treatment, the presence of appropriate raw materials is crucial to produce oxidants.

Their impact on the effectiveness of EAOPs is vital as it allows for the spread of toxin oxidation from the anode's surface to the whole electrolyte. It is crucial to acknowledge that these oxidants have a significant influence on the processes of contaminant oxidation. In certain cases, they may also result in the creation of unwanted intermediate or final stable products. At times, the entities that produce oxidants are not naturally found in the wastewater but are intentionally introduced as reagents. This results in the achievement of very efficient protocols. An important use of Ag (II) is in the treatment of garbage, where it has been demonstrated to be highly effective when used with conductive electrodes (Figure 7.1).

An often-mentioned instance of assisted electrochemical oxidation is the influence of chlorides on the oxidation of organic molecules. Chlorides are often found in most wastewater flow channels and may be easily transformed into chlorine by oxidation using several types of anode materials. Permeation enables the ingress of the gaseous oxidant into the wastewater, where it undergoes a chemical reaction, leading to the production of hypochlorite and chloride. Hypochlorous acid can undergo deprotonation, resulting in the formation of hypochlorite (Martinez-Huitle & Ferro, 2006). It is commonly noted in literature that the conversion of chloride to hypochlorite is often described directly, without mentioning the entire sequence of reactions.

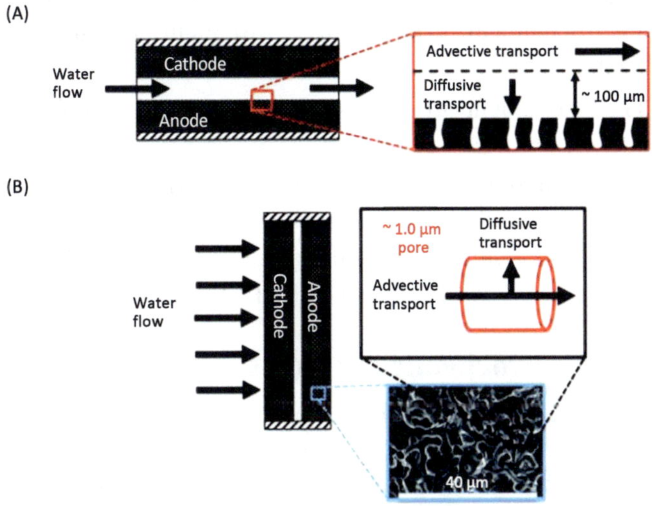

Figure 7.1. *Development of stable electrode materials (A) flow-by electrode operation; (B) flow-through electrode operation.*

Source: Chaplin, Creative Commons License.

Nevertheless, oxidation in that environment is conducted by a combination of chemicals, where the specific amount of each component is contingent upon the concentration and pH.

$$2Cl^- \rightarrow Cl_2 + 2e^-$$

$$Cl_2 + H_2O \rightarrow HClO + H^+ + Cl^- \quad \text{Acidic}$$

$$Cl_2 + 2\,OH^- \leftrightarrow ClO^- + Cl^- + H_2O \quad \text{Alkaline}$$

$$HClO + OH^- \leftrightarrow ClO^- + H_2O$$

$$Cl^- + H_2O \rightarrow ClO^- + 2H^+ + 2e^-$$

The resulting mixture is extremely reactive with various organic compounds, leading to their mineralization. Nevertheless, it is acknowledged that the creation of different organo-chlorinated compounds may take place as intermediates and end products, posing a considerably greater risk than the initial toxin. Dealing with these species can be quite demanding, especially when it comes to handling lightweight substances that have the potential to cause unconsciousness when inhaled. Proper organization and precautions are of utmost importance (Chaplin, 2014). While chlorine-mediated oxidation is well-known, it is just one example of mediated oxidation processes and may not be the most significant. When focusing on the progress of mediated oxidation, it is important to consider two key aspects:

- The anode surface can produce oxidants directly through electrochemical means using non-oxidant substances found in the waste.
- The movement of these substances to the main body of the wastewater.

The presence of the raw material necessary to produce oxidants in the wastewater or its controlled addition is essential. Usually, this substance can exist as a particle (like chloride or sulfate), a toxic chemical (like acetic acid), dissolved gases (like oxygen), or even water (Radha & Sirisha, 2018).

7.1.1. Photo-Electrochemical Processes

The use of photoelectrochemical methods for the treatment of water and wastewater has seen a sharp rise in attention. These medications harness the power of a compromised system or a light-sensitive electrode to UV or solar light sources. UV-A, UV-B, and UV-C lights emitted by UV lights are widely utilized as energy sources. The intensity and frequency of these radiations greatly influence the rate at which organic pollutants are broken down (Georgieva et al., 2012). While UV-assisted techniques may come with a higher price tag, their utilization provides notable benefits. One way to achieve this is by using solar-powered systems, which capture sunlight to provide a cost-effective and eco-friendly energy solution. Occasionally, the effects of this radiation are assessed using a solar test device. This text offers a comprehensive overview of various photoelectrochemical forms, such as solar photoelectro-Fenton (SPEF), photoelectro-Fenton (PEF), photoelectrocatalysis (PEC), and hybrid systems. It focuses on their distinct features and primary applications (Kim et al., 2020).

Did You Know?

Electrochemical oxidation processes have a wide range of applications, including wastewater treatment and the synthesis of valuable chemicals and materials. This highlights their range of uses and potential for sustainable industrial use.

7.1.2. Photoelectro-Fenton (PEF) and Solar Photoelectro-Fenton (SPEF) Processes

The Brillas' group has conducted extensive research on the Photoelectro-Fenton (PEF) process, utilizing simulated UV-A light. They have also explored the Solar Photoelectro-Fenton (SPEF) process, which harnesses daylight. These procedures include treating the contaminated system under certain conditions while exposing it to UV-A or solar radiation to speed up the mineralization of organic molecules. The Fenton's reaction produces hydroxyl radicals, which aid in further oxidation (Brillas, 2014). To ensure the treatment procedure remains efficient, it is essential to prevent the development of insoluble Fe(III) particles and the consequent decrease of $[Fe(OH)]^{2+}$ by photolysis. The Fe(III) species depicted in Figure 7.2 exhibit a discernible arrangement within a pH range of 2.8–3.5. This arrangement serves the purpose of replenishing the catalyst Fe^{2+} in Fenton's reaction and generating more radicals (Figure 7.3). Furthermore, radiation can facilitate the breakdown of certain oxidation intermediates or their Fe(III) complexes by photolysis, resulting in the regeneration of Fe^{2+} in a manner akin to photodecarboxylation processes (Clematis & Panizza, 2021).

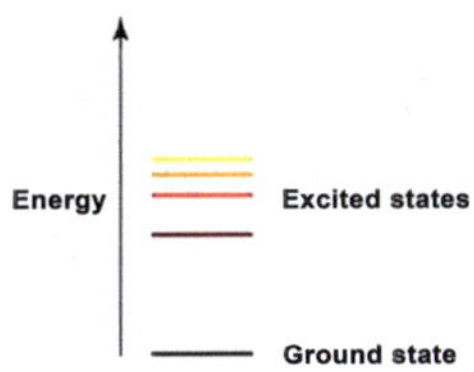

Figure 7.2. *Illustration of electron jump from ground state to a higher energy excited state.*

Source: Suresh Ameta, Creative Commons License.

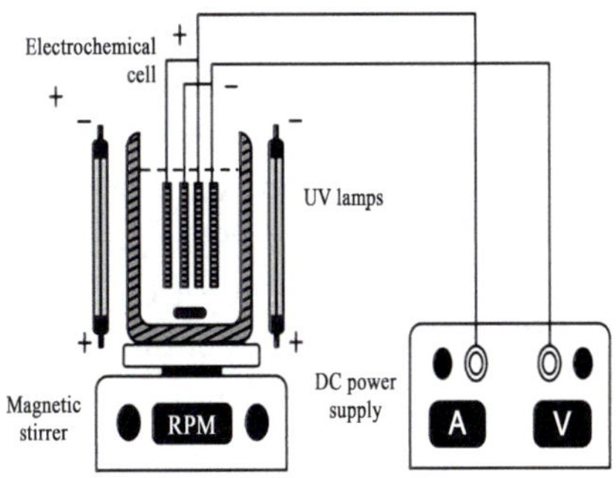

Figure 7.3. *Schematic representation of photoelectron-Fenton process.*

Source: Rakshit Ameta, Creative Commons License.

The assessment of the SPEF technique has been expanded to a 10-liter pre-pilot plant. The photoreactor of this plant consists of a cathode zone with a surface area of 90.3 cm^2, which is connected to a solar compound parabolic collector (CPCs) with a volume of 1.57 L. Extensive study and experiments have conclusively shown that the most effective method to obtain a substantial 75% decrease in total organic carbon (TOC) and an excellent current efficiency of 71% is by applying specified conditions. They include a 5 A electric current, a 0.4 mM concentration of Fe^{2+}, and the maintenance of a pH level of 3.0. It is important to mention that this procedure utilizes 93 kilowatt-hours (kWh) of energy for every kilogram of total organic carbon (TOC). The experimental circumstances involved testing a paracetamol solution with a concentration of 157 mg/L, in the presence of 0.05 M Na_2SO_4, over a duration of 120 minutes (Nidheesh et al., 2023).

An investigation was undertaken using High Performance Liquid Chromatography (HPLC) to detect hydroquinone and other benzoquinones as aromatic intermediates. The hydroxyl radicals (•OH) successfully removed these intermediates. In addition, carboxylic acids such as fumaric, maleic, succinic, oxalic, formic, lactic, and oxamic acids were detected. In a recent study, researchers employed a particular approach called SPEF. The procedure required the utilization of an exact concentration of 297 mg/L of the azo hue sunset yellow FCF. The treatment solution consisted of a 0.05 M concentration of Na_2SO_4 and a 0.5 mM concentration of Fe^{2+} at a pH of 3.0. The experiment was likewise carried out at a pH of 7.0. It was demonstrated that complete discoloration may be accomplished within a time frame of 120 minutes. A remarkable rate of mineralization of around 94% was attained with a vitality use of 197 kWh (kg TOC)$^{-1}$ within a mere 150 minutes (Fajardo et al., 2019).

The study done by IsarainChavez et al. (2011) intended to examine the corruption of SPEF in setups including metoprolol tartrate, atenolol, and propranolol hydrochloride, with a concentration of 100 mg/L TOC. The studies were carried out in a solution with a concentration of 0.10 M Na_2SO_4 and 0.5 mM Fe^{2+} at a pH of 3.0. The researchers employed single Pt/adenine and BDD/adenine cells, as well as a combination of Pt/adenine and Pt/CF cells, to enhance the extraction of Fe^{2+} from Fe^{3+} reduction. The processes demonstrate the configuration of the integrated BDD/Adenine-Pt/CF cell. The investigation revealed that the utilization of a 0.246 mM metoprolol tartrate solution resulted in a significant augmentation in the interconnection of cells, as compared to their independent existence. In addition, BDD electrodes have shown more performance in comparison to Pt electrodes, whereas SPEF has proven higher efficacy over EF in reducing total organic carbon (TOC) levels. This phenomenon is caused by the higher production of hydroxyl radicals (•OH) through Fenton's reaction in the combined cells, the strong oxidizing ability of boron-doped diamond (BDD) (•OH), and the capacity of sunlight to initiate photolysis in solar photoelectro-Fenton (SPEF), which can be easily achieved using a series of cells. Nevertheless, the Pt/ADE-Pt/CF cell exhibited an impressive energy consumption of 80 kWh (kg TOC)$^{-1}$ for 88%–93% mineralization, establishing it as a very cost-effective technology for industrial applications (Figure 7.4) (Khataee et al., 2010).

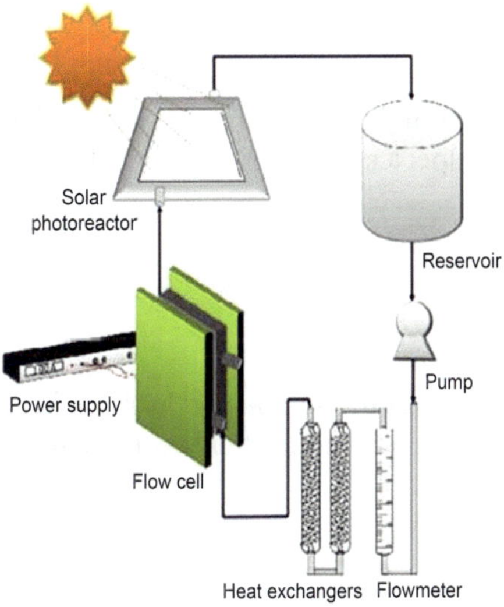

Figure 7.4. *Diagram of solar photoelectron-Fenton system.*

Source: Trellu, Creative Commons License.

7.1.3. Photoelectrocatalysis (PEC)

The PEC technique employs the integration of electrolysis and photocatalysis to augment the efficacy of wastewater cleanup. The utilization of the nanocrystalline anatase form of TiO_2 in photocatalysis has been widely researched for the purpose of oxidizing organic contaminants in water under light exposure.

This semiconductor material has desirable properties like cost-effectiveness, little toxicity, and a broad energy band gap of 3.2 eV. These characteristics enhance its exceptional longevity and inhibit photocorrosion.

An electron is elevated from the valence band to the conduction band (e^{-CB}), when anatase TiO_2 nanoparticles, either in a colloidal suspension or as a thin film on Ti, are exposed to UV photons with energy greater than or equal to 380 nm. This results in the formation of a positively charged vacancy (hole) in the valence band (h^{+VB}).

$$TiO + hv \rightarrow e_{CB}^{-} + h_{VB}^{+}$$

Organic molecules can undergo selective oxidation, either by the action of the hole or the heterogeneous hydroxyl radical. This process takes place when water molecules are adsorbed and react with the photogenerated species (Zhang et al., 2023).

$$h_{VB}^{+} + H_2O \rightarrow {}^{\bullet}OH + H^{+}$$

Furthermore, the photoinduced electron possesses the capacity to generate diverse reactive oxygen species (ROS), such as •OH, via specialized reactions.

$$e_{CB}^{-} + O_2 \rightarrow O_2^{\bullet -}$$

$$O_2^{\bullet -} + H^{+} \rightarrow HO_2^{\bullet}$$

$$2HO_2^{\bullet} \rightarrow H_2O_2 + O_2$$

$$H_2O_2 + O_2^{\bullet -} \rightarrow {}^{\bullet}OH + OH^{-} + O_2$$

One of the key reasons for the significant decrease in efficiency is primarily due to the combination of photoinduced electrons with either unreacted holes or adsorbed •OH.

$$e_{CB}^{-} + h_{VB}^{+} \rightarrow TiO_2 + heat$$

$$e_{CB}^{-} + {}^{\bullet}OH \rightarrow OH^{-}$$

The PEC process is based on applying a continuous anodic voltage to a UV-exposed thin film semiconductor-based anode.

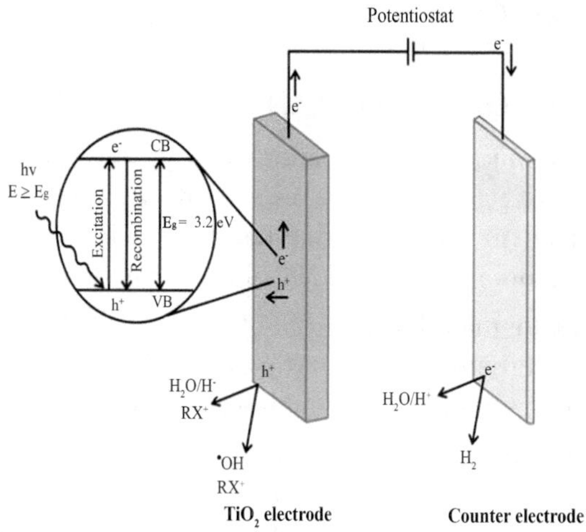

Figure 7.5. *Enhancement of photoelectrocatalysis efficiency using nanostructured electrodes.*

Source: Suresh Ameta, Creative Commons License.

As a result, photoinduced electrons can continue to be extracted using an external electrical circuit. PEC electrolytic cells utilize either mixed tanks or flow reactors to directly expose the solution or photoanode surface to UV light. This is achieved through a quartz window, which helps minimize the loss of incident illumination (Figure 7.5) (Cheng et al., 2017).

7.1.4. Hybrid Combinations of PEF and PEC

Researchers made advancements in the removal of color from some dyes by employing a combination of Pulsed Electric Fields (PEF) and Titanium Dioxide (TiO_2) photocatalysis (TiO_2/UV). They improved the experimental conditions by implementing a reaction surface approach. The developers employed open, integrated, and cylindrical mixed tank reactors with a capacity of either 1 or 3 liters. The reactors were furnished with a platinum anode and a cathode made up of carbon nanotubes incorporated into a graphite surface (CNT/graphite). The reactors were supplied with air and exposed to a 6 W UV-C light source enclosed in a quartz tube. In addition, the researchers utilized paper or glass plates to immobilize TiO_2 nanoparticles, effectively applying a layer of coating on the inner surface of the reactor. The decolorization rate of azo color acid yellow 36 decreased in the following sequence (Liu et al., 2019):

$$\text{PEF-TiO}_2/\text{UV} > \text{PEF} > \text{EF} > \text{TiO}_2/\text{UV} > \text{UV photolysis}$$

The best parameters for the prior combined technique were determined to be 0.15 mM Fe^{3+}, 25 mg/L of dye, 127 minutes of treatment, and 115.6 mA. The utilization of these parameters led to an impressive 83% efficiency in removing color. The operating costs associated with the use of UV light reached a maximum of 16.5 kWh/m^3, but the electricity usage was only 0.88 kWh/m^3. However, researchers have undertaken many endeavors to enhance the oxidation capacity of PEC by combining it with EF. Nevertheless, further investigation is necessary to confirm the effectiveness of these combined versions (Divyapriya et al., 2022).

Xie & Li (2006) employed a quartz reactor to successfully eliminate the azo dye orange G using several techniques over a period of 5 hours. A 0.1 mM solution of the dye, together with 0.01 M Na_2SO_4 at pH 6.2, was added to the cell, filling it with a volume of 30 mL. A UV-A light source with a power output of 8 watts was used for the experiment. UV-A light exposure did not result in any discernible photolysis of Orange G. However, a little reduction (about 3%) was seen when the TiO_2/Pt cell was treated with AO at an anodic potential of +0.71 V/SCE. In addition, TiO_2 photocatalysis led to an 8% decrease by producing small quantities of oxidizing agents (•OH and holes) on the surface of TiO_2 (Oriol et al., 2019). The use of the PEC technique with the Pt/TiO_2 cell resulted in a significant 25% improvement in dye removal. This phenomenon can be explained by the increased formation of oxidizing holes by photogeneration. The oxidation efficiency of this process greatly increased when the Pt cathode was replaced with a reticulated vitreous carbon (RVC) anode at an Ecat value of –0.54 V/SCE. This modification allowed for the oxidation of the dye by using hydrogen peroxide (H_2O_2) produced by the reduction of oxygen (O_2). It is significant to note that the orange G totally disappeared when the PEF was applied using a Fe/RVC cell with about 17 mg/L Fe^{2+}, a pH level of 3.0, and Ecat set at –0.71 V/SCE. Furthermore, there was a marginal augmentation in the pace of discoloration when EF was amalgamated with PEC utilizing a TiO_2/RVC cell (Azhdeh et al., 2023). This outcome was a consequence of an augmented generation of uniform •OH radicals by Fenton's reaction, in addition to the additional stress induced by the Fenton reaction. In research done by Peralta-Herna´ndez et al. in 2006, the combination of PEC and EF resulted in a mineralization rate of 74% during a span of only 5 hours. The study demonstrated a significant improvement in the effectiveness of decolorization and removal of total organic carbon (TOC) by using a combination of photoelectrochemical (PEC) and electrocoagulation-flotation (EF), as compared to using EF alone. The objective was accomplished by utilizing a concentric annular integrated TiO_2/graphite flow cell, together with a central UV light source with an intensity of 75 mW/cm^2 (Oturan et al., 2008).

Remember

Addressing challenges such as high energy consumption, electrode fouling, and scalability limitations is crucial for the widespread adoption of electrochemical oxidation technologies.

7.1.5. Sonoelectrochemical Processes

Pyrolysis can utilize ultrasonic (US) to generate highly reactive radicals in water, which in turn aids in the decomposition of natural poisons present. However, its ability to undergo oxidation is somewhat restricted. Therefore, it is frequently employed in combination with other oxidizing agents such as O_2, UV, H_2O_2, and Fenton's reagent to efficiently treat water. The current areas of active research focus on investigating the integration of US and terminal operations, as well as examining the utilization of sonoelectrochemical technology in the combustion of organic materials and flow dynamics. This section focuses on the features of sono-electrolysis and sonoelectro-Fenton, two significant sonoelectrochemical techniques employed for the purpose of sterilizing wastewater (Tran et al., 2017) (Figure 7.6).

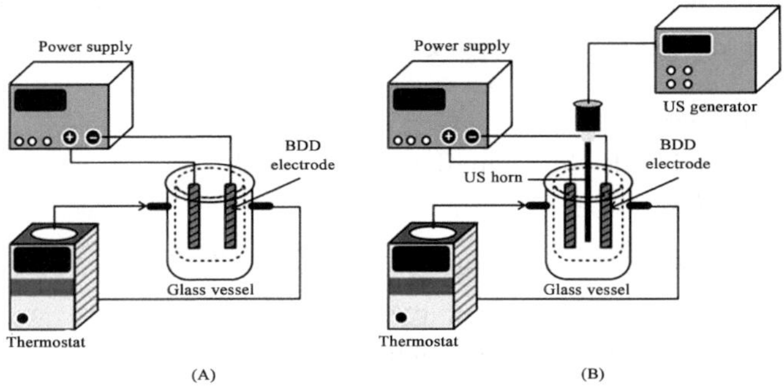

Figure 7.6. *Schematic representation of (A) electrochemical cell; and (B) sonoelectrochemical cell.*

Source: Suresh Ameta, Creative Commons License.

Practice Problem

Let's explore the benefits and obstacles of combining photoelectro-Fenton (PEF) and sonoelectrochemical processes in a hybrid electrochemical oxidation reactor for wastewater treatment.

Solution to Practice Problem

The integration of PEF and sonoelectrochemical processes in a hybrid reactor offers several advantages. PEF methods employ both photoelectrochemical and Fenton reactions, using renewable energy sources to decompose contaminants. Acoustic cavitation enhances mass transfer and reaction kinetics in sonoelectrochemical processes, leading to increased degradation rates of contaminants. By combining these processes, the hybrid reactor can generate impressive synergistic effects, leading to a substantial improvement in treatment efficiency. However, there are other obstacles that need to be addressed in order to effectively execute the hybrid system, such as the intricacy of reactor design, the optimization of operating conditions, and the possible problems related to scale. To get the best treatment performance, it is essential to establish compatibility between the two processes and optimize the setup of the hybrid reactor.

7.2. PROS AND CONS OF ELECTROCHEMICAL OXIDATION PROCESSES

Learning Objectives:

- Comprehend the benefits and drawbacks of electrochemical oxidation processes.

There are several advantages to these types of processes:

- Utilizing electrons as a reagent provides a clean and effective solution for addressing contamination issues.
- It exhibits an exceptional level of energy efficiency.
- Its automation is a straightforward process.
- The device is user-friendly, reliable, and adaptable.
- It is applicable to effluents with COD levels ranging from 0.1 g/L to 100 g/L (Rhandi et al., 2020).

There are certain limitations associated with these types of processes:

- The cost becomes excessively high when it is linked to the provision of electricity.
- Some wastewaters have low conductance, so electrolytes need to be added.
- One downside to consider is the decline in activity and shortened lifespan of the electrode due to fouling. This occurs when organic material is deposited on its surface (Kamali et al., 2023).

7.3. APPLICATIONS OF ELECTROCHEMICAL OXIDATION PROCESSES

Learning Objectives:

- Understand the applications of electrochemical oxidation processes on the surface of electrode.
- Comprehend the role of electrochemical oxidation processes for the treatment of landfill leachates.

1. **Applications in Wastewater Treatment:** In recent decades, electrochemical techniques have become a prominent method for addressing environmental cleanup. An essential strategy is the control and handling of liquid waste generated by many sectors. Electrolysis and electro-coagulation have gained popularity as

pre-treatment techniques in recent years, demonstrating significant potential as sophisticated technologies. This technique possesses the capacity to eliminate turbidity, de-colorize water, and disintegrate waste materials that include emulsions. The electrolysis technique has difficulties in removing large-scale contamination in industrial waste due to the substantial energy requirements needed to eradicate contaminants present in high quantities (Anglada et al., 2009). Electrolysis is frequently used to treat wastewater that contains artificial organic chemicals. This treatment is particularly efficient for effluents with a concentration of chemical oxygen demand (COD) ranging from 1,000 to 20,000 milligrams per cubic decimeter (mg/dm^{-3}) during direct anodic oxidations. Conductive-diamond electrochemical oxidation in electrolysis exhibits some remarkable characteristics in comparison to other advanced oxidation techniques. With its phenomenal capabilities, it can effortlessly decompose a wide range of organic substances, leaving behind no stubborn residues. It operates with high efficiency, achieving near-perfect efficiencies when there are no diffusion limitations. Furthermore, it seamlessly integrates with various treatment technologies and can be efficiently powered by renewable energy sources like photovoltaic panels (Qiao & Xiong, 2021).

2. **Applications on Surface of Electrodes:** EAOPs may be classified into two primary types depending on the processes that govern the oxidation process: surface oxidation and bulk oxidation occurring within the electrochemical cell. Surface-controlled processes, sometimes referred to as anodic oxidation processes, may be categorized into three classes according to the oxidation mechanisms. The occurrence of these processes is mostly concentrated in the vicinity of the electrode's surface (Shestakova & Sillanpää, 2017).

3. In 2013, Klauck et al. showcased the use of electrochemical oxidation to treat landfill leachates and evaluate the toxicity in allium cepa. This study sought to evaluate the effectiveness of electrochemical oxidation, a cutting-edge oxidation process, in the treatment of leachate. In addition, the sample's toxicity was assessed before and after treatment. Although the dye was effectively removed during the treatment process, the removal of chlorine odor did not meet expectations. The presence of a large amount of chlorine ions in the leachate enables them to function as oxidizing agents during the indirect electrochemical oxidation of organic substances. The process was shown to be effective in removing the physico-chemical characteristics. Nevertheless, the eco-toxicological test revealed that

the effluent exhibited a comparable degree of toxicity to the original sample (Szpyrkowicz, 2006).

Further optimizing the parameters, including current, time and pH is required. It is important to consider implementing integrated treatment systems, as well as utilizing ozone, photo- and electro-oxidation processes. In addition, it is crucial to thoroughly examine the leachate following biological treatment, carry out experiments using alternative biomarkers, and investigate the production of organic compounds and by-products (Tallman & Petersen, 1990).

7.4. PROSPECTS OF ELECTROCHEMICAL OXIDATION PROCESSES

Learning Objectives:

- Comprehend future developments and research directions in electrochemical oxidation processes for wastewater treatment.

- The potential for enhancing interceded oxidation in future EAOP advancements appears to be promising. Interceded oxidation plays a vital role in enhancing the efficiency of EAOPs. Exploring the initiation of oxidants in reaction media is a captivating area of research, which surpasses the mere production of oxidants (Sirés et al., 2014).
- The costs of innovation can be quite significant, and it is not always clear if the improvements in efficiency justify their integration with EAOPs. This can be attributed to the significant levels of energy spread out across this technology. Curiosity about this topic should arise from a more strategic approach to energy utilization, aiming to address a range of challenges through potentially bold measures.
- It is interesting to note that light illumination is currently showing great promise as a highly favorable alternative with excellent prospects for future use. It has been clearly demonstrated that activating oxidants has a combined effect, resulting in a significantly lower energy output compared to the input (Pan et al., 2019).
- The use of the distinctive BDD anode greatly improved the oxidation capacity and efficiency of mineralization by producing a higher number of hydroxyl radicals on the anode's surface. Therefore, this innovative development has now reached a point where it may be efficiently employed in the construction of a pilot-scale reactor for treating large amounts of wastewater. The first stage will necessitate

the creation and organization of a pilot-scale reactor (Radjenovic & Sedlak, 2015).

- Combining the treatment with an organic procedure, whether it is done before or after, shows great potential for achieving a cost-effective treatment. Recent reviews have indicated that it is indeed possible to achieve such a combination.
- The EF procedure can efficiently convert both harmful and bio-refractory particles into biodegradable species in a short treatment period. Organic treatment can lead to complete mineralization.
- By analyzing the potential outcomes of combining SPEC with SPEF in PEC proposals, one can anticipate the expected progress. This untested hybrid system is anticipated to offer a higher level of cost-effectiveness compared to current PEC combinations (Mandal et al., 2017).

SUMMARY

- The chapter explores the significance of electrochemical oxidation methods in wastewater treatment, emphasizing their contribution to addressing pollution concerns and advancing environmental sustainability.
- It provides a comprehensive overview of different electrochemical oxidation processes. It covers topics such as photoelectron-Fenton (PEF), photoelectrochemical, solar photoelectron-Fenton, hybrid combination of PEF and PEC, photoelectrocatalysis (PEC), and sonoelectrochemical processes.
- Moreover, this chapter provides a comprehensive evaluation of the benefits and limitations of electrochemical oxidation techniques. It explores different factors such as energy usage, treatment effectiveness and scalability, and operational costs, providing valuable insights into the suitability of electrochemical oxidation processes in various wastewater treatment situations.
- Finally, the chapter delves into the practical applications of electrochemical oxidation processes in various industries and facilities for the treatment of wastewater. It also explores the future advancements and research directions in the field of electrochemical oxidation processes.

REVIEW QUESTIONS

1. What are the fundamental principles underlying photoelectrochemical processes, and how do they contribute to pollutant degradation in wastewater treatment?
2. Discuss the role of semiconductor materials in photoelectrocatalysis (PEC) and its significance in electrochemical oxidation processes for wastewater treatment.
3. How do hybrid electrochemical oxidation reactors integrate different processes to enhance treatment efficiency? Provide examples and potential applications.
4. How do sonoelectrochemical processes combine ultrasound and electrochemistry for pollutant degradation in wastewater treatment? Discuss the mechanisms involved and potential benefits.
5. Explore the real-world applications of electrochemical oxidation processes in various industries and wastewater treatment facilities. What are some notable case studies and success stories?

MULTIPLE CHOICE QUESTIONS

1. **Which process involves the generation of electron-hole pairs under illumination to drive oxidation reactions?**
   a. Sonoelectrochemical processes
   b. Photoelectro-Fenton (PEF) processes
   c. Hybrid combination of PEF and PEC
   d. Photoelectrochemical processes
2. **What is the primary objective of Solar Photoelectro-Fenton (PEF) processes?**
   a. Minimizing energy consumption
   b. Maximizing reactor scalability
   c. Utilizing renewable energy for pollutant degradation
   d. Enhancing pollutant adsorption efficiency
3. **Which process involves the use of photoactive electrodes to catalyze oxidation reactions under illumination?**
   a. Sonoelectrochemical processes
   b. Photoelectrocatalysis (PEC)
   c. Hybrid combination of PEF and PEC
   d. None of above
4. **What is the key advantage of hybrid electrochemical oxidation reactors?**
   a. Reduced treatment efficiency
   b. Increased operational costs
   c. Enhanced treatment efficiency through synergistic effects
   d. Decreased scalability
5. **Which combination of processes integrates ultrasound and electrochemistry for wastewater treatment?**
   a. Hybrid combination of PEF and PEC
   b. Photoelectro-Fenton (PEF) processes
   c. Sonoelectrochemical processes
   d. None of above

Answers to Multiple Questions

1. (d); 2. (c); 3. (b); 4. (c); 5. (c)

REFERENCES

1. Anglada, A., Urtiaga, A., & Ortiz, I. (2009). Contributions of electrochemical oxidation to wastewater treatment: Fundamentals and review of applications. *Journal of Chemical Technology & Biotechnology, 84*(12), 1747–1755. https://doi.org/10.1002/jctb.2214.
2. Azhdeh, A., Mashhadizadeh, M. H., & Moazami, H. R. (2023). Developing a photo-electric-field wireless electrochemical system for highly efficient removal of diazinon as an organic model pollutant as a next-generation electrochemical advanced oxidation process. *Journal of Applied Electrochemistry, 53*(6), 1219–1243. https://doi.org/10.1007/s10800-022-01695-4.
3. Barrera-Díaz, C., Canizares, P., Fernández, F. J., Natividad, R., & Rodrigo, M. A. (2014). Electrochemical advanced oxidation processes: An overview of the current applications to actual industrial effluents. *Journal of the Mexican Chemical Society, 58*(3), 256–275.
4. Brillas, E. (2014). A review on the degradation of organic pollutants in waters by UV photoelectro-Fenton and solar photoelectro-Fenton. *Journal of the Brazilian Chemical Society, 25*(3), 393–417. https://doi.org/10.5935/0103-5053.20140028.
5. Canizares, P., Paz, R., Sáez, C., & Rodrigo, M. A. (2009). Costs of the electrochemical oxidation of wastewaters: A comparison with ozonation and Fenton oxidation processes. *Journal of Environmental Management, 90*(1), 410–420. https://doi.org/10.1016/j.jenvman.2007.10.010.
6. Chaplin, B. P. (2014). Critical review of electrochemical advanced oxidation processes for water treatment applications. *Environmental Science: Processes & Impacts, 16*(6), 1182–1203. https://doi.org/10.1039/c3em00679d.
7. Cheng, L., Liu, L., Yan, K., & Zhang, J. (2017). Visible light-driven photo electrocatalysis coupling with electroenzymatic process for degradation of chloramphenicol. *Chemical Engineering Journal, 330*, 1380–1389. https://doi.org/10.1016/j.cej.2017.08.004.
8. Clematis, D., & Panizza, M. (2021). Electro-Fenton, solar photoelectro-Fenton and UVA photoelectro-Fenton: Degradation of Erythrosine B dye solution. *Chemosphere, 270*, 129480. https://doi.org/10.1016/j.chemosphere.2021.129480.
9. Darvishmotevalli, M., Zarei, A., Moradnia, M., Noorisepehr, M., & Mohammadi, H. (2019). Optimization of saline wastewater treatment using electrochemical oxidation process: Prediction by RSM method. *MethodsX, 6*, 1101–1113. https://doi.org/10.1016/j.mex.2019.05.021.

10. Divyapriya, G., Srinivasan, R., Mohanalakshmi, J., & Nambi, I. M. (2022). Development of a hybrid bifunctional rotating drum electrode system for the enhanced oxidation of ciprofloxacin: An integrated photo electrocatalysis and photo-electro-Fenton processes. *Journal of Water Process Engineering, 49*, 102967. https://doi.org/10.1016/j.jwpe.2022.102967.

11. Fajardo, A. S., dos Santos, A. J., de Araújo Costa, E. C., da Silva, D. R., & Martínez-Huitle, C. A. (2019). Effect of anodic materials on solar photoelectro-Fenton process using a diazo dye as a model contaminant. *Chemosphere, 225*, 880–889. https://doi.org/10.1016/j.chemosphere.2019.03.088.

12. Georgieva, J., Valova, E., Armyanov, S., Philippidis, N., Poulios, I., & Sotiropoulos, S. (2012). Bi-component semiconductor oxide photoanodes for the photo electrocatalytic oxidation of organic solutes and vapors: A short review with emphasis to TiO_2–WO_3 photoanodes. *Journal of Hazardous Materials, 211*, 30–46. https://doi.org/10.1016/j.jhazmat.2011.10.055.

13. Kamali, M., Aminabhavi, T. M., Costa, M. E. V., Ul Islam, S., Appels, L., & Dewil, R. (2023). Heterogeneous Advanced Oxidation Processes (HE-AOPs) for the Removal of Pharmaceutically Active Compounds—Pros and Cons. In *Advanced Wastewater Treatment Technologies for the Removal of Pharmaceutically Active Compounds* (Vol. 3, pp. 211–239). Cham: Springer International Publishing. https://doi.org/10.1007/978-3-030-70582-7_10.

14. Khataee, A. R., Zarei, M., & Asl, S. K. (2010). Photocatalytic treatment of a dye solution using immobilized TiO2 nanoparticles combined with photoelectro-Fenton process: Optimization of operational parameters. *Journal of Electroanalytical Chemistry, 648*(2), 143–150. https://doi.org/10.1016/j.jelechem.2010.08.005.

15. Kim, D. S., Kim, Y. B., Jung, S. H., Deshpande, N. G., Choi, J. H., Lee, H. S., & Cho, H. K. (2020). Atomically tunable photo-assisted electrochemical oxidation process design for the decoration of ultimate-thin CuO on Cu_2O photocathodes and their enhanced photoelectrochemical performances. *Journal of Materials Chemistry A, 8*(41), 21744–21755. https://doi.org/10.1039/D0TA07888A.

16. Liu, C. F., Huang, C. P., Hu, C. C., & Huang, C. (2019). A dual TiO_2/Ti-stainless steel anode for the degradation of orange G in a coupling photoelectrochemical and photo-electro-Fenton system. *Science of The Total Environment, 659*, 221–229. https://doi.org/10.1016/j.scitotenv.2018.12.380.

17. Mandal, P., Dubey, B. K., & Gupta, A. K. (2017). Review on landfill leachate treatment by electrochemical oxidation: Drawbacks, challenges and future scope. *Waste Management, 69*, 250–273. https://doi.org/10.1016/j.wasman.2017.08.034.
18. Martinez-Huitle, C. A., & Ferro, S. (2006). Electrochemical oxidation of organic pollutants for the wastewater treatment: Direct and indirect processes. *Chemical Society Reviews, 35*(12), 1324–1340. https://doi.org/10.1039/B517632H.
19. Moreira, F. C., Boaventura, R. A., Brillas, E., & Vilar, V. J. (2017). Electrochemical advanced oxidation processes: A review on their application to synthetic and real wastewaters. *Applied Catalysis B: Environmental, 202*, 217–261. https://doi.org/10.1016/j.apcatb.2016.08.037.
20. Nidheesh, P. V., Trellu, C., Vargas, H. O., Mousset, E., Ganiyu, S. O., & Oturan, M. A. (2023). Electro-Fenton process in combination with other advanced oxidation processes: Challenges and opportunities. *Current Opinion in Electrochemistry, 37*, 101171. https://doi.org/10.1016/j.coelec.2022.101171.
21. Oriol, R., Sirés, I., Brillas, E., & De Andrade, A. R. (2019). A hybrid photo electrocatalytic/photoelectro-Fenton treatment of Indigo Carmine in acidic aqueous solution using TiO_2 nanotube arrays as photoanode. *Journal of Electroanalytical Chemistry, 847*, 113088. https://doi.org/10.1016/j.jelechem.2019.113088.
22. Oturan, M. A., Sirés, I., Oturan, N., Pérocheau, S., Laborde, J. L., & Trévin, S. (2008). Sonoelectro-Fenton process: A novel hybrid technique for the destruction of organic pollutants in water. *Journal of Electroanalytical Chemistry, 624*(1–2), 329–332. https://doi.org/10.1016/j.jelechem.2008.09.011.
23. Pan, Z., Song, C., Li, L., Wang, H., Pan, Y., Wang, C., & Feng, X. (2019). Membrane technology coupled with electrochemical advanced oxidation processes for organic wastewater treatment: Recent advances and future prospects. *Chemical Engineering Journal, 376*, 120909. https://doi.org/10.1016/j.cej.2019.120909.
24. Qiao, J., & Xiong, Y. (2021). Electrochemical oxidation technology: A review of its application in high-efficiency treatment of wastewater containing persistent organic pollutants. *Journal of Water Process Engineering, 44*, 102308. https://doi.org/10.1016/j.jwpe.2021.102308.
25. Radha, K. V., & Sirisha, K. (2018). Electrochemical oxidation processes. In *Advanced Oxidation Processes for Wastewater Treatment* (Vol. 2, pp. 359–373). Academic Press. https://doi.org/10.1016/B978-0-12-805302-8.00012-8.

26. Radjenovic, J., & Sedlak, D. L. (2015). Challenges and opportunities for electrochemical processes as next-generation technologies for the treatment of contaminated water. *Environmental Science & Technology, 49*(19), 11292–11302. https://doi.org/10.1021/acs.est.5b02414.

27. Rhandi, M., Trégaro, M., Druart, F., Deseure, J., & Chatenet, M. (2020). Electrochemical hydrogen compression and purification versus competing technologies: Part I. Pros and cons. *Chinese Journal of Catalysis, 41*(5), 756–769. https://doi.org/10.1016/S1872-2067(19)63526-6.

28. Shestakova, M., & Sillanpää, M. (2017). Electrode materials used for electrochemical oxidation of organic compounds in wastewater. *Reviews in Environmental Science and Bio/Technology, 16*, 223–238. https://doi.org/10.1007/s11157-017-9432-5.

29. Sirés, I., Brillas, E., Oturan, M. A., Rodrigo, M. A., & Panizza, M. (2014). Electrochemical advanced oxidation processes: Today and tomorrow. A review. *Environmental Science and Pollution Research, 21*, 8336–8367. https://doi.org/10.1007/s11356-014-2783-1.

30. Szpyrkowicz, L. (2006). Application of electrochemical oxidation for treatment of industrial wastewater—The influence of reactor hydrodynamics on direct and mediated processes. *Journal of Chemical Technology & Biotechnology: International Research in Process, Environmental & Clean Technology, 81*(8), 1375–1383. https://doi.org/10.1002/jctb.1546.

31. Tallman, D. E., & Petersen, S. L. (1990). Composite electrodes for electroanalysis: Principles and applications. *Electroanalysis, 2*(7), 499–510. https://doi.org/10.1002/elan.1140020702.

32. Tran, N., Drogui, P., Brar, S. K., & De Coninck, A. (2017). Synergistic effects of ultrasounds in the Sono electrochemical oxidation of pharmaceutical carbamazepine pollutant. *Ultrasonics Sonochemistry, 34*, 380–388. https://doi.org/10.1016/j.ultsonch.2016.06.036.

33. Wu, Z., & Zhou, M. (2001). Partial degradation of phenol by advanced electrochemical oxidation process. *Environmental Science & Technology, 35*(13), 2698–2703. https://doi.org/10.1021/es001649y.

34. Zhang, X., Yu, W., Guo, Y., Li, S., Chen, Y., Wang, H., & Bian, Z. (2023). Recent advances in photo electrocatalytic advanced oxidation processes: From mechanism understanding to catalyst design and actual applications. *Chemical Engineering Journal, 455*, 140801. https://doi.org/10.1016/j.cej.2023.140801.

35. Zhou, M., & He, J. (2007). Degradation of azo dye by three clean advanced oxidation processes: Wet oxidation, electrochemical oxidation and wet electrochemical oxidation—A comparative study. *Electrochimica Acta, 53*(4), 1902–1910. https://doi.org/10.1016/j.electacta.2007.07.071.

CHAPTER 8

APPLICATIONS OF UV-BASED AOPS IN WASTEWATER TREATMENT

LEARNING OBJECTIVES

At the end of this chapter, readers will be able to:

1. Understand the fundamentals of UV-based advanced oxidation techniques for treating wastewater.
2. Identify the different types of UV-based AOPs, including UV/H2O2, UV/Ozone, and UV/Fenton processes, and their respective mechanisms.
3. Comprehend how UV radiation interacts with ozone, hydrogen peroxide, or Fenton reagents to generate reactive oxygen species for pollutant degradation.
4. Comprehend the applications of UV-based AOPs in wastewater treatment.

INTRODUCTORY EXAMPLE

CleanWater Solutions Inc. recently collaborated with a municipal wastewater treatment plant facing challenges in meeting stringent effluent quality standards. Traditional treatment methods were proving inadequate in removing persistent organic pollutants and pathogens, leading to concerns about water quality and environmental impact.

To address these challenges, CleanWater Solutions proposed the implementation of UV-based AOPs, leveraging the powerful oxidative capabilities of ultraviolet (UV) radiation in combination with hydrogen peroxide or ozone. By harnessing the synergistic effects of UV irradiation and advanced oxidation reactions, CleanWater Solutions aimed to enhance the degradation of organic contaminants and disinfection of microbial pathogens in the wastewater.

Through a pilot-scale demonstration at the wastewater treatment plant, CleanWater Solutions showcased the efficacy of UV-based AOPs in achieving significant reductions in pollutant concentrations and microbial load. The successful implementation of UV-based AOPs not only improved effluent quality but also demonstrated the potential for innovative water treatment technologies to address complex pollution challenges.

As CleanWater Solutions continues to innovate and optimize UV-based AOPs for wastewater treatment applications, the company remains committed to providing cost-effective, sustainable solutions that safeguard water resources and promote environmental stewardship.

UNIT INTRODUCTION

There has been a rise in the presence of rigid and difficult-to-remove pollutants in water and wastewater, prompting the development of different technologies, one of which is advanced oxidation processes (AOPs). These processes shown significant potential in the treatment of a wide range of pollutants across different concentrations. They have been successfully used in different water treatment applications, including wastewater, groundwater, and landfill leachate (Ibrahim et al., 2019).

Photochemical technology is commonly utilized in advanced oxidation processes. In recent years, there has been a growing interest in these technologies because of their dual benefits of contaminant treatment and disinfection. In addition to that, the simple and efficient operation is another reason why this technology is chosen for water and wastewater treatment. UV irradiation is commonly employed in AOPs due to its ability to boost the effectiveness of oxidizing agents, resulting in the formation of free radicals. These oxidizing agents include hydrogen peroxide (H_2O_2), ozone (O_3) and chlorine (Zhang et al., 2022).

Photochemical technology is frequently used in advanced oxidation processes. These technologies have gained significant popularity in recent years because of their dual benefits of contaminant disinfection and treatment. In addition to that, simple and efficient operation is another reason why this technology is chosen for water and wastewater treatment. UV irradiation is frequently utilized in AOPs due to its ability to boost the effectiveness of oxidizing agents, resulting in the formation of free radicals. These oxidizing agents include hydrogen peroxide (H_2O_2), ozone (O_3) and chlorine (Yang et al., 2014).

In addition, oxidation of organic pollutants results in the formation of carbon dioxide, water, and mineral acids, as well as simpler, less refractory organic molecules (Crittenden et al., 2012). Strong oxidizing species like the hydroxyl radical (OH•) are essentially produced by ultraviolet (UV) based radical AOPs through photo-Fenton processes, heterogeneous photocatalysis or direct photolysis of hydrogen peroxide (H_2O_2). The main process involved in UV direct photolysis is the degradation caused by the absorption of UV light radiation (Sgroi et al., 2021).

Thus, the emphasis is typically placed on the pollutants that have a high affinity for absorbing UV radiation. Furthermore, the combination of UV direct photolysis with H_2O_2 generates OH radicals, which effectively eliminate contaminants through oxidation processes. Therefore, water contaminants can be effectively treated using a combination of UV radiation and processes induced by OH radicals. Many UV light absorbers consist of double bonds or conjugated double bonds with carbon, nitrogen, or oxygen atoms. These compounds are known for their delocalized π-electrons (Miklos et al., 2018).

There has been a significant increase in the use of UV-light for water and wastewater treatment due to concerns about the presence of potentially harmful contaminants and the strict regulations imposed by water authorities. As a result, numerous studies have been carried out to investigate the degradation of various organic and inorganic compounds that are of environmental concern, induced by UV-light (Buthiyappan et al., 2015).

UV-based advanced oxidation processes, such as photocatalysis reactions (photo-Fenton, UV/TiO_2) and photooxidation (UV/O_3, UV/H_2O_2, UV/O_3/H_2O_2) have been found to exhibit much higher rates of pollutant removal compared to direct photolysis. This chapter will provide a thorough examination of the different UV-based radical AOPs. It delves into the underlying principles and fundamentals of these processes, while also exploring their current applications and potential future opportunities for treatment (Deng & Zhao, 2015).

Remember

UV-based advanced oxidation processes (AOPs) harness the power of UV radiation to generate reactive oxygen species, such as hydroxyl radicals, for wastewater treatment.

8.1. UV RADIATION WITH H_2O_2

Learning Objectives:

- Comprehend the principle of UV/H2O2 advances oxidation process.
- Understand the photolysis of hydrogen peroxide.
- Comprehend the hydrogen peroxide as a UV radiation disinfection method.
- Identify the applications of hydrogen peroxide with UV.

The fundamental concept of AOP revolves around the generation of hydroxyl radicals (OH●), that can be produced by combining hydrogen peroxide ($H_2O_{2)}$ with ultraviolet (UV) radiation. This method efficiently breaks down organic compounds and gets rid of harmful microbes in water-based solutions by combining H_2O_2 and UV radiation. The hydroxyl radical's generation is crucial to the success of this process, as they play a significant role in its effectiveness (Benitez et al., 2013).

8.1.1. Photolysis of H_2O_2 to Form Free Radical Species

When UV light at a wavelength of 254 nm is applied to water, it causes the photolytic dissociation of H_2O_2, which results in the generation of hydroxyl radicals. The H_2O_2/UV process generates hydroxyl radicals by breaking down hydrogen peroxide and initiating subsequent reactions. One commonly accepted mechanism for the photolysis of H_2O_2 is represented by the equation below (Varanasi et al., 2018).

When UV light with a wavelength of 254 nm is used to photolytically dissociate H_2O_2 in water, hydroxyl radicals are produced. The process of H_2O_2/UV generates hydroxyl radicals through the photolysis of hydrogen peroxide and subsequent propagation reactions. One commonly accepted mechanism for the photolysis of H_2O_2 is presented by the equation below (Keen & Linden, 2013).

$H_2O_2 + h\nu \rightarrow 2\ HO^\bullet\ (\lambda \leq 254\ nm)$

UV radiation is represented by hv.

The most common way to produce UV radiation is to use low-pressure mercury vapor UV lamps, which peak at 254 nm. In their study, Cooper & Zika (1983) discussed the potent oxidizing properties of H_2O_2. They observed that H_2O_2 naturally breaks down into water and oxygen, a process that is facilitated by sunlight in natural surface water. This decomposition is influenced by the presence of natural organic elements (Chuang et al., 2017).

Therefore, the process of direct photolysis of H_2O_2 results in the generation of hydroxyl (OH) radicals. For a considerable period, H_2O_2 has been employed to break down organic substances in both industrial and municipal wastewater. Additionally, it can be utilized for the sterilization of wastewater treatment facilities. The combination of H_2O_2 with UV radiation, catalysts, or other oxidants significantly enhances the efficacy of the treatment process, leading to a more efficient destruction of organic compounds found in high strength wastewaters.

Therefore, the direct photolysis of H_2O_2 results in the generation of OH radicals. For a long time, H_2O_2 has been used to break down organic compounds in both industrial and municipal wastewater. Moreover, it can be used to disinfect wastewater treatment facilities. The treatment procedure is improved and becomes much more successful when H_2O_2 is combined with UV radiation, catalyst, or other oxidants to get rid of the organic components present in concentrated wastewaters.

The combination of a catalyst with H_2O_2 can lead to the generation of a diverse range of free radicals and reactive species, which can transform or break down organic chemicals. Although the chemistry underlying H_2O_2 reactions is complicated, it may be divided into several fundamental concepts and reaction mechanisms. These involve various processes, including free radical chain reactions (Pandis et al., 2022):

- Steps involving initiation, propagation, and termination;
- Eliminating radicals;
- Breaking down organic contaminants;
- Balancing reactions and minimizing non-productive processes.

According to Collivignarelli et al. (2017), the effectiveness of the UV/H_2O_2 process is influenced by several factors that impact its capability to break down organic molecules. These variables include the kind and quantity of dissolved inorganic materials (such as carbonates and metallic cations) or organic contaminants, the pH, temperature, light-transparent quality of the solutions, and the quantity of hydrogen peroxide applied.

When a free radical is initiated by photolysis, a series of easy reactions will take place. The complexity of the chemistry can be assigned to the sheer number of reactions involved. Therefore, accurately predicting all the oxidation products can be quite challenging due to the complex mechanism involved. Several factors can influence the concentration of radicals, including (Rodríguez, 2003):

- pH;
- Temperature;
- Ion's presence;
- Pollutant type;
- Scavengers' presence (bicarbonate ion).

When employing the UV/H_2O_2 approach, it is crucial to highlight the synergistic effect of the UV radiation and H_2O_2, which is advantageous only with pollutants that require rather severe oxidation conditions (high activation energies) (Wang et al., 2022).

8.1.2. Decontamination Mechanism

The process by which UV/H_2O_2 decontaminates UV radiation includes the given subsections.

8.1.2.1. H_2O_2 Oxidation

The process of moving electrons from a reductant to an oxidant that has a greater affinity for electrons is known as oxidation. The transfer of electrons leads to the chemical conversion of both the oxidizing agent and the reducing agent, sometimes resulting in the development of chemical species with an uneven amount of valence electrons, referred to as radicals, which tend to be highly unstable (Azizi et al., 2017).

Therefore, radicals show higher reactivity due to the presence of an unpaired electron. The process of oxidizing water contaminants results in the conversion of organic elements in the contaminants into simple, comparatively inorganic, and harmless molecules, a process known as mineralization. At the same time, the reactions that cause oxidation and produce radicals have a tendency to occur, leading to additional oxidation events with various reactants (both inorganic and organic). This process persists until oxidation products that are thermodynamically stable are produced (Collivignarelli et al., 2017). The oxidation potential of an oxidant is a measure of its ability to trigger chemical reactions. Fluorine, ozone, hydroxyl radicals (OH●), and chlorine are the most potent oxidants. The term "radical" can also be represented as OH● or OH.' Therefore, the most effective concentration of H_2O_2 relies on the quantity of contaminants that need to be oxidized. The hydroxyl radical (OH●) functions as a potent oxidizing agent that can eliminate various organic and inorganic molecules present in water. OH● radicals are characterized by their non-selective reactivity, allowing them to react with a diverse array of pollutants without the need for additional additives (Yin et al., 2021).

Furthermore, the OH● species can interact with organic molecules in two distinct manners: by eliminating a hydrogen atom or by introducing a hydrogen atom to double bonds. As a result, new oxidized intermediates are produced, which have a lower molecular weight. When complete mineralization takes place, the resulting end products are water and carbon dioxide (Bein et al., 2023).

8.1.2.2. UV Photolysis

The production of hydroxyl radicals occurs through the photolysis of hydrogen peroxide, resulting in the efficient oxidation of the targeted compounds. Photolysis is a process where compounds absorb photons, leading to the release of energy that triggers oxidation reactions when exposed to light. Determining the photolysis rate of a compound involves considering multiple factors, such as its ability to absorb light and its quantum yield. The extinction coefficient shows the correlation between wavelength and photon absorption, indicating that as the wavelength decreases, a higher number of photons are absorbed.

The hydroxyl radical's generation, which drives the UV/H_2O_2 process, is facilitated by the photolysis of hydrogen peroxide. Using UV light, a powerful oxidizing agent known as hydroxyl radical (OH•) is produced. This radical is recognized as one of the most potent oxidants, exceeding the effectiveness of hydrogen peroxide alone (Amin et al., 2008). As a result, it exhibits superior efficiency in eliminating microorganisms and breaking down organic substances in water (Guo et al., 2020).

$$H_2O_2 + \text{UV light} \dashrightarrow 2\,OH^\bullet$$

In general, organic substances tend to be more light-sensitive compared to inorganic substances. The process of degradation can happen when harsh UV photons are directly absorbed or when water photolysis products, including OH• react with them. The radicals play a crucial role in the breakdown of dissolved organic matter into carbon dioxide and water, contributing to its mineralization (Yu et al., 2016).

8.1.3. Application of H_2O_2 with UV

Because of its strong oxidizing ability, H_2O_2 is an excellent choice for getting rid of a variety of contaminants, such as germs, hazardous organic compounds, and certain metals. Numerous uses for the procedure exist in the manufacture of drinking water, including regulating taste and odor, getting rid of metals, boosting ozone levels, and sanitizing the water. Combining UV with hydrogen peroxide (indirect photolysis) can generate a highly reactive and unselective H_2O_2, that has demonstrated effectiveness in eliminating certain organic pollutants (Dhawle et al., 2021).

Did You Know?

UV-based AOPs can be integrated into existing wastewater treatment plants as a tertiary treatment step, enhancing overall treatment efficiency and water quality improvement.

8.2. UV RADIATION WITH OZONE

Learning Objectives:

- Understand the principle of UV radiation with ozone.
- Comprehend the decontamination mechanism and process of generating oxidative species.
- Identify the applications of ozone with UV.

UV photons have been examined in conjunction with ozone, as they can activate ozone molecules. Through the generation of additional hydroxyl radicals within the treatment system, the utilization of ozone and UV radiation is expected to enhance the effectiveness of the ozonation process. The O_3/UV approach can be used to successfully oxidize and eliminate organic molecules that are toxic and water-refractory. Below is a simple reaction sequence (Bernhard et al., 2023).

$$O_3 + H_2O \rightarrow O_2 + H_2O_2 \text{ (in the presence of UV light)}$$

$$2\,O_3 + H_2O_2 \rightarrow 2\,OH^\bullet + 3\,O_2$$

8.2.1. Oxidative Species Generation

The majority of OH• is created from ozone through a chain reaction in which hydrogen peroxide (H_2O_2) or OH^- (high pH) serve as starters. A key element in the O_3/UV combination's effective destruction of organic contaminants is the hydroxyl radical's high oxidation potential.

According to Brandon et al. (1997), the primary reactive species responsible for the breakdown of organic toxicants is the free radical created during ozone breakdown (Gurol & Vatistas, 1987).

When ozone is excited in aqueous solution, it combines with water to generate OH•|, a chemical oxidant that is less selective and stronger than either one alone. This process is known as O_3/UV reactions, or photolytic ozonation. In meanwhile, ozone efficiently combines with organic pollutants to form two hydroxyl radicals per ozone molecule, signaling oxidation targets to many attackers and significantly altering the oxidation environment (Rosenfeldt et al., 2006).

The extremely nonselective and active hydroxyls act on a wide range of targets, increasing the efficiency with which the ozone molecule's function. Therefore, UV light wavelengths between 250 and 260 nm are responsible for the destruction of ozone residuals, while UV light wavelengths between 100 and 400 nm are responsible for the inactivation of microorganisms, with a wavelength of 254 nm being the most effective. Direct photolysis, molecular ozone, and OH• radical reactions all lead to organic oxidation (Pell et al., 1997).

8.2.2. Mechanism of the Decontamination

The decontamination mechanism involving UV radiation and ozone (UV/O_3) includes UV photolysis. UV photolysis is an interesting procedure where compounds absorb photons, leading to the release of energy that triggers oxidation reactions under the influence of light. Quantum yield and the compound's rate of light absorption can be used to estimate a compound's photolysis rate (Peyton et al., 1982).

Photolysis alone can effectively reduce certain organic compounds, like N-nitrosodimethylamine (NDMA). Several crucial factors contribute to the effectiveness of the UV/O_3 system, including the dosage of ozone, the level of UV irradiation, and the pH level. Ensuring a high rate of dissolved ozone as well as effective transfer of ozone gas into the aqueous solution is essential for achieving accurate ozone dosing (Epelle et al., 022).

During the process of ozonation, the oxidation of organic and inorganic compounds can take place in two ways. One way is through molecular ozone, which is the direct reaction that is typically observed in an acidic environment. The other way is through hydroxyl radicals, which is the indirect reaction that is more commonly seen in an alkaline environment (Brodowska et al., 2018).

8.2.3. Application of UV/O_3

Several studies have explored the effectiveness and speed of oxidation on various surfaces through the use of combined treatment involving UV radiation and ozone. These studies have found that the combination treatment, which includes additional ozone generated by the UV radiation, resulted in more efficient and rapid oxidation compared to treatments involving ozone alone or UV treatment alone. Combining UV radiation with ozone leads to a significantly enhanced and accelerated oxidation of polypropylene, polyethylene surfaces and polyethylene terephthalate, surpassing the effectiveness of ozone or UV treatment in isolation. Various treatment methods involve different reaction mechanisms (Liu et al., 2021).

Remember

AOPs have applications beyond wastewater treatment, including air purification and disinfection of surfaces and drinking water.

8.3. PHOTO FENTON PROCESS ($H_2O_2/FE^{3+}/UV$)

Learning Objectives:

- Understand the mechanism of photo Fenton process.
- Comprehend the application of photo Fenton process.

The Fenton method was investigated using Fenton's reagent, which is a combination of ferrous iron and hydrogen peroxide (H_2O_2). This combination generates hydroxyl radicals (Fenton, 1894). Enhancement of the mineralization extent in the decomposition of H_2O_2 can be achieved by utilizing ferrous iron (Fe^{2+}) or ferric iron (Fe^{3+}) in combination with H_2O_2. These reagents can be further optimized by subjecting them to UV radiation and visible light irradiation.

The ferrous iron (Fe^{2+}) initiates and accelerates the breakdown of H_2O_2, leading to the generation of hydroxyl radicals. The increased production of hydroxyl radicals (OH•) due to the combined action of oxidant chemicals and metallic catalysts in the presence of UV radiation, along with the possibility of utilizing sunlight as a UV light source, renders this system extremely attractive.

Furthermore, the utilization of hydrogen peroxide and UV radiation in conjunction with Fe^{2+} or Fe^{3+} oxalate ion (referred to as the photo-Fenton (PF) process) has been found to generate a higher concentration of hydroxyl radicals when compared to traditional Fenton methods or photolysis. This enhanced production of hydroxyl radicals leads to an accelerated degradation rate of organic pollutants. In an innovative investigation, Fenton (1894) showed the remarkable oxidizing capabilities of a combination of H_2O_2 and Fe^{2+} in an acidic environment. A simple redox reaction is used in the pure mechanism, wherein H_2O_2 is reduced to the hydroxide ion and the hydroxyl radical and Fe^{2+} is oxidized to Fe^{3+} (Kuo et al., 2012).

$$Fe^{2+} + H_2O_2 \rightarrow Fe^{3+} + OH\bullet + OH^-$$

Furthermore, the efficiency of the reaction can be enhanced using UV/V irradiation, known as the Fenton reaction. This process generates extra OH radicals and facilitates the regeneration of the Fe^{2+} catalysts. This reaction serves as the main driver of photochemical self-cleaning in the atmosphere.

$$Fe^{3+} + H_2O + hv \rightarrow Fe^{2+} + H^+ + OH\bullet$$

These reactions demonstrate a recurring cycle of iron transitioning between Fe^{2+} and Fe^{3+} states. The system's overall stoichiometry indicates the generation of two moles of OH• for every mole of H_2O_2 used. In a study conducted by Jr. et al. (2010), it was noted that the acceleration of the Fenton reaction can be achieved by any reaction or process that enhances the conversion of Fe^{3+} to Fe^{2+}.

In a study they discussed the use of UV irradiation in combination with the traditional Fenton system to improve the reduction of dissolved Fe^{3+} to Fe^{2+}.

The concentration of hydrogen peroxide is a key factor in determining the efficiency of the degradation process. It is widely recognized that the breakdown of pollutants tends to be more effective as the quantity of H_2O_2 increases.

The total effectiveness of the degrading process is largely dependent on the concentration of hydrogen peroxide. Research has shown that the pollutant degradation percentage tends to rise as the dosage of hydrogen peroxide is increased. However, it is important to beware when choosing the dosage of the oxidant for operation. It is not advisable to have any excess amount of unused hydrogen peroxide during the Fenton process, as it can contribute to COD. In addition, the inclusion of hydrogen peroxide has been found to have detrimental effects on various organisms, thereby significantly impacting the overall efficiency of degradation. This is particularly relevant when Fenton oxidation is utilized as a preliminary step to biological oxidation. In a study, researchers provided a rationale for the control of the Fenton reaction process, specifically focusing on the levels of Fe^{2+}, H_2O_2, and pH. The effectiveness of the Fenton process is greatly influenced by the pH of the solution, primarily because of the speciation factors of iron and hydrogen peroxide. It was discovered that the Fenton reaction achieves optimal results at a pH of approximately 3, regardless of the specific substrate being targeted (Chu et al., 2006). Increasing the pH level decreased the effectiveness of the Fenton reagent in reacting with less reactive iron oxohydroxides and forming a precipitate of ferric hydroxide. This circumstance suggests that the presence of fewer free iron ions leads to a decrease in the generation of hydroxyl radicals. The oxidation potential of hydroxyl radicals is inversely proportional to the pH level. The rate of degradation increases as the concentration of iron ions increases. However, the extent of increase is sometimes minimal over a particular concentration. An abundance of ferrous ions might hinder the photochemical degradation process because Fe^{2+} may compete with organic compounds for oxidant agents (hydroxyl radicals), resulting in inhibition (Bag et al., 2009).

8.3.1. Application of Photo-Fenton Process ($H_2O_2/Fe^{3+}/UV$)

Exploring Fenton and photo Fenton processes have proven to be effective in treating various types of wastewaters, such as those generated in pulp bleaching, dye manufacturing, agricultural processing, and chemical manufacturing. Various industries can benefit from the application of Photo-Fenton processes to treat their wastewater. These industries include the landfill leachate, plastic industry, municipal wastewater, dye house industry, pesticides, organic effluent, phenolic wastes, and the paper industry (Xu, 2001) (Table 8.1).

Table 8.1. *Representation of Advantages and Disadvantages of UV-based AOPs*

AOP Process	Advantages	Disadvantages
H_2O_2/UV	• UV irradiation can be used as a disinfectant. • There are already full-scale drinking water treatment systems in place. • H_2O_2/UV treatment is effective in mineralizing organic pollutants. It has been found to be more efficient than the Photo-Fenton process in removing nonylphenol polyethoxylate surfactant and reducing total organic carbon (TOC) levels.	• UV light penetration can be affected by turbidity. • Generates OH• less efficiently than the O_3/UV process in terms of stoichiometry. • The UV absorption characteristics of H_2O_2 are not ideal. • If the water matrix has a high UV light absorption capacity, a significant amount of the light energy input to the reactor will not be utilized efficiently.

O_3/UV	• Exceptional performance in water treatment. • Superior effectiveness compared to O_3 and UV. It has been observed that the efficiency of generating OH is higher in the advanced oxidation process compared to the H_2O_2/UV process, even when the oxidant concentrations are equal. • Action of disinfection.	• It requires a significant amount of energy and can be quite expensive. • The presence of turbidity can hinder the penetration of UV light. • Diffusion of ozone may lead to limitations in mass transfer.
Photo Fenton $H_2O_2/Fe^{3+}/UV$	• Environmentally friendly to water ecosystems. • Regenerates the consumed Fe^{2+} ions during illumination. • Converts toxic compounds into non-toxic or less hazardous substances. • Demonstrates high efficiency and cost-effectiveness.	• Generates a significant concentration of anions in the wastewater being treated. • Possibility of generating substantial quantities of ferrous iron sludge. • Requires the presence of acidic conditions.

Source: Abdul Aziz, Creative Commons License.

Practice Problem

Compare and contrast the UV/Ozone and UV/Fenton processes for wastewater treatment, highlighting their respective advantages and limitations.

Solution to Practice Problem

The UV/Ozone process involves the photolysis of ozone (O_3) molecules by UV radiation, generating reactive oxygen species such as hydroxyl radicals (•OH) and singlet oxygen. These species contribute to the oxidative degradation of pollutants in wastewater. In contrast, the UV/Fenton process combines UV radiation with a Fenton-like reagent, typically consisting of iron salts and hydrogen peroxide. Under UV irradiation, the Fenton reagent generates hydroxyl radicals through the photodecomposition of hydrogen peroxide and the photoreduction of iron ions. While both processes are effective in degrading organic pollutants, the UV/Ozone process requires the addition of ozone, which may increase operational costs. On the other hand, the UV/Fenton process may involve the use of iron salts, which can lead to secondary pollution if not properly managed.

SUMMARY

- Advanced oxidation processes are a highly effective method for treating water and wastewater. They utilize various reacting systems to break down stubborn organic compounds and eliminate harmful microorganisms.
- UV-based AOPs have gained significant recognition for their ability to effectively destroy target micro-pollutants and inactivate pathogens. One of the key aspects of UV-based AOPs is the generation of OH radicals. This oxidant is a great choice for eliminating a variety of difficult-to-remove contaminants because of its remarkable reactivity and non-selectivity nature. In a similar vein, OH radicals are essential for the inactivation of infections via microbial activities. The formula is true even when the source of the disinfectant changes.
- Applying AOPs remains a significant challenge due to the complexity of pollutants in potable water and wastewater resulting from advancements in chemical industries. Developing systems for purifying huge amounts of water is made more difficult by the diversity of chemicals and substances found in water.
- The combination of UV radiation and hydrogen peroxide (UV/H_2O_2) results in the generation of hydroxyl radicals by breaking down hydrogen peroxide using UV light. This process improves the breakdown of organic pollutants, pharmaceutical residues, and microorganisms in wastewater, providing a highly effective and eco-friendly treatment choice.
- The combination of UV radiation and ozone (UV/O_3) employs UV irradiation to break down ozone (O_3) molecules, resulting in the production of reactive oxygen species (ROS) like hydroxyl radicals and singlet oxygen. The presence of an oxidative environment enables the breakdown of organic pollutants, colorants, and compounds that cause unpleasant odors in wastewater, leading to enhanced water quality and visual appeal.
- The photo-Fenton process combines UV radiation with a Fenton-like reagent, usually made up of iron salts (such as ferrous sulfate) and hydrogen peroxide. When exposed to UV irradiation, the Fenton reagent produces hydroxyl radicals by breaking down hydrogen peroxide and reducing iron ions through a photochemical process. In this advanced oxidation process, a highly oxidative environment is created to effectively break down stubborn organic pollutants found in wastewater. These pollutants can include industrial dyes, pesticides, and pharmaceuticals.

REVIEW QUESTIONS

1. Explain the principles underlying UV-based Advanced Oxidation Processes (AOPs) in wastewater treatment. How do these processes utilize ultraviolet (UV) radiation to degrade contaminants?
2. Compare and contrast the UV/H_2O_2, UV/Ozone, and UV/Fenton processes for wastewater treatment. What are the differences in their mechanisms and applications?
3. Discuss the factors influencing the efficiency of UV-based AOPs in degrading contaminants in wastewater?
4. Provide examples of organic pollutants commonly degraded by UV-based AOPs in wastewater treatment.
5. Explore the role of reactive oxygen species generated in UV-based AOPs, such as hydroxyl radicals and singlet oxygen, in degrading pollutants in wastewater.
6. Evaluate the advantages and limitations of UV-based AOPs compared to conventional wastewater treatment methods.

MULTIPLE CHOICE QUESTIONS

1. Which UV-based Advanced Oxidation Process (AOP) involves the photolysis of hydrogen peroxide by ultraviolet irradiation to generate hydroxyl radicals?

a. UV/Ozone

b. UV/H_2O_2

c. UV/Fenton

d. UV/Chlorine

2. What is the primary role of UV radiation in UV/Ozone process for wastewater treatment?

a. To generate ozone molecules

b. To activate the catalyst

c. To photolyze ozone and generate reactive oxygen species

d. To neutralize microbial contaminants

3. Which UV-based AOP involves the combination of UV radiation with a Fenton-like reagent?

a. UV/Ozone

b. UV/H_2O_2

c. UV/Fenton

d. UV/Chlorine

4. What reactive oxygen species are typically generated in UV/H_2O_2 process for wastewater treatment?

a. Singlet oxygen

b. Hydroxyl radicals

c. Superoxide ions

d. All of the above

5. What is the purpose of combining UV radiation with ozone in UV/Ozone process for wastewater treatment?

a. To increase UV penetration depth

b. To reduce UV energy consumption

c. To generate more reactive oxygen species

d. To neutralize alkalinity in wastewater

6. Which of the following pollutants is efficiently degraded by UV-based AOPs?

a. Inorganic salts

b. Heavy metals

c. Organic pollutants

d. Suspended solids

7. Which UV-based AOP is particularly effective for degrading recalcitrant organic pollutants?

a. UV/Ozone

b. UV/H_2O_2

c. UV/Fenton

d. UV/Chlorine

Answers to the Multiple Questions

1. (b); 2. (c); 3. (c); 4. (b); 5. (c); 6. (c); 7. (c)

REFERENCES

1. Azizi, E., Fazlzadeh, M., Ghayebzadeh, M., Hemati, L., Beikmohammadi, M., Ghaffari, H. R., & Sharafi, K. (2017). Application of advanced oxidation process (H_2O_2/UV) for removal of organic materials from pharmaceutical industry effluent. *Environment Protection Engineering,*

43(1), 183–191. https://doi.org/10.5277/epe170115.

2. Bag, B. C., Sai, M., Sekhar, K., & Bhattacharya, C. (2009). Treatment of wastewater containing pyridine released from N, N'-Dichlorobis (2, 4, 6-trichlorophenyl) urea (CC_2) plant by advanced oxidation. *Journal of Environmental Protection Science, 3*, 34–40.
3. Bein, E., Seiwert, B., Reemtsma, T., Drewes, J. E., & Hübner, U. (2023). Advanced oxidation processes for removal of monocyclic aromatic hydrocarbon from water: Effects of O_3/H_2O_2 and UV/H_2O_2 treatment on product formation and biological post-treatment. *Journal of Hazardous Materials, 450*, 131066. https://doi.org/10.1016/j.jhazmat.2023.131066.
4. Benitez, F. J., Acero, J. L., Real, F. J., Roldan, G., & Rodriguez, E. (2013). Modeling the photodegradation of emerging contaminants in waters by UV radiation and UV/H_2O_2 system. *Journal of Environmental Science and Health, Part A, 48*(1), 120–128. https://doi.org/10.1080/10934529.2013.726267.
5. Bernhard, G. H., Bais, A. F., Aucamp, P. J., Klekociuk, A. R., Liley, J. B., & McKenzie, R. L. (2023). Stratospheric ozone, UV radiation, and climate interactions. *Photochemical & Photobiological Sciences, 22*(5), 937–989. https://doi.org/10.1007/s43630-022-00082-w.
6. Brodowska, A. J., Nowak, A., & Śmigielski, K. (2018). Ozone in the food industry: Principles of ozone treatment, mechanisms of action, and applications: An overview. *Critical Reviews in Food Science and Nutrition, 58*(13), 2176–2201. https://doi.org/10.1080/10408398.2017.1308313.
7. Buthiyappan, A., Abdul Aziz, A. R., & Wan Daud, W. M. A. (2015). Degradation performance and cost implication of UV-integrated advanced oxidation processes for wastewater treatments. *Reviews in Chemical Engineering, 31*(3), 263–302. https://doi.org/10.1515/revce-2014-0033.
8. Chu, W., Kwan, C. Y., Chan, K. H., & Wu, C. W. (2006). Comprehensive study of the wavelength effect on oxidation processes of 2-chloroaniline. *Industrial & Engineering Chemistry Research, 45*(11), 3769–3775. https://doi.org/10.1021/ie060173y.
9. Chuang, Y. H., Chen, S., Chinn, C. J., & Mitch, W. A. (2017). Comparing the UV/monochloramine and UV/free chlorine advanced oxidation processes (AOPs) to the UV/hydrogen peroxide AOP under scenarios relevant to potable reuse. *Environmental Science & Technology, 51*(23), 13859–13868. https://doi.org/10.1021/acs.est.7b03414.
10. Collivignarelli, M. C., Pedrazzani, R., Sorlini, S., Abbà, A., & Bertanza, G. (2017). H2O2 based oxidation processes for the treatment of real high

strength aqueous wastes. *Sustainability, 9*(2), 244. https://doi.org/10.3390/su9020244.

11. Deng, Y., & Zhao, R. (2015). Advanced oxidation processes (AOPs) in wastewater treatment. *Current Pollution Reports, 1*, 167–176. https://doi.org/10.1007/s40726-015-0015-z.
12. Dhawle, R., Frontistis, Z., Mantzavinos, D., & Lianos, P. (2021). Production of hydrogen peroxide with a photocatalytic fuel cell and its application to UV/H_2O_2 degradation of dyes. *Chemical Engineering Journal Advances, 6*, 100109. https://doi.org/10.1016/j.ceja.2021.100109.
13. Epelle, E. I., Macfarlane, A., Cusack, M., Burns, A., Mackay, W. G., Rateb, M. E., & Yaseen, M. (2022). Application of ultraviolet-c radiation and gaseous ozone for microbial inactivation on different materials. *ACS Omega, 7*(47), 43006–43021. https://doi.org/10.1021/acsomega.2c05545.
14. Guo, K., Wu, Z., & Fang, J. (2020). UV-based advanced oxidation process for the treatment of pharmaceuticals and personal care products. In *Contaminants of Emerging Concern in Water and Wastewater* (Vol. 3, pp. 367–408). Butterworth-Heinemann. https://doi.org/10.1016/B978-0-12-813564-8.00013-5.
15. Gurol, M. D., & Vatistas, R. (1987). Oxidation of phenolic compounds by ozone and ozone + UV radiation: A comparative study. *Water Research, 21*(8), 895–900. https://doi.org/10.1016/0043-1354(87)90074-8.
16. Ibrahim, N., Zainal, S. F. F. S., & Aziz, H. A. (2019). Application of UV-based advanced oxidation processes in water and wastewater treatment. In *Advanced Oxidation Processes (AOPs) in Water and Wastewater Treatment* (Vol. 3, pp. 384–414). IGI Global. https://doi.org/10.4018/978-1-5225-7214-5.ch013.
17. Keen, O. S., & Linden, K. G. (2013). Degradation of antibiotic activity during UV/H2O2 advanced oxidation and photolysis in wastewater effluent. *Environmental Science & Technology, 47*(22), 13020–13030. https://doi.org/10.1021/es402457b.
18. Kuo, C. Y., Pai, C. Y., Wu, C. H., & Jian, M. Y. (2012). Effects of oxidant concentration and temperature on decolorization of azo dye: Comparisons of UV/Fenton and UV/Fenton-like systems. *Water Science and Technology, 65*(11), 1970–1974. https://doi.org/10.2166/wst.2012.106.
19. Liu, H., Gao, Y., Wang, J., Ma, D., Wang, Y., Gao, B., & Xu, X. (2021). The application of UV/O_3 process on ciprofloxacin wastewater containing high salinity: Performance and its degradation mechanism. *Chemosphere, 276*, 130220. https://doi.org/10.1016/j.chemosphere.2021.130220.
20. Miklos, D. B., Remy, C., Jekel, M., Linden, K. G., Drewes, J. E., &

Hübner, U. (2018). Evaluation of advanced oxidation processes for water and wastewater treatment–A critical review. *Water Research, 139*, 118–131. https://doi.org/10.1016/j.watres.2018.03.042.

21. Pandis, P. K., Kalogirou, C., Kanellou, E., Vaitsis, C., Savvidou, M. G., Sourkouni, G., & Argirusis, C. (2022). Key points of advanced oxidation processes (AOPs) for wastewater, organic pollutants and pharmaceutical waste treatment: A mini review. *ChemEngineering, 6*(1), 8. https://doi.org/10.3390/chemengineering6010008.
22. Pell, E. J., Schlagnhaufer, C. D., & Arteca, R. N. (1997). Ozone-induced oxidative stress: Mechanisms of action and reaction. *Physiologia Plantarum, 100*(2), 264–273. https://doi.org/10.1111/j.1399-3054.1997.tb04787.x.
23. Peyton, G. R., Huang, F. Y., Burleson, J. L., & Glaze, W. H. (1982). Destruction of pollutants in water with ozone in combination with ultraviolet radiation. 1. General principles and oxidation of tetrachloroethylene. *Environmental Science & Technology, 16*(8), 448–453. https://doi.org/10.1021/es00103a005.
24. Rodríguez, M. (2003). Fenton and UV-vis based advanced oxidation processes in wastewater treatment: Degradation, mineralization and biodegradability enhancement. *Universitat de Barcelona. 8*(4), 87–98.
25. Rosenfeldt, E. J., Linden, K. G., Canonica, S., & Von Gunten, U. (2006). Comparison of the efficiency of OH radical formation during ozonation and the advanced oxidation processes O_3/H_2O_2 and UV/H_2O_2. *Water Research, 40*(20), 3695–3704. https://doi.org/10.1016/j.watres.2006.09.008.
26. Sgroi, M., Anumol, T., Vagliasindi, F. G., Snyder, S. A., & Roccaro, P. (2021). Comparison of the new $Cl_2/O_3/UV$ process with different ozone- and UV-based AOPs for wastewater treatment at pilot scale: Removal of pharmaceuticals and changes in fluorescing organic matter. *Science of the Total Environment, 765*, 142720. https://doi.org/10.1016/j.scitotenv.2020.142720.
27. Varanasi, L., Coscarelli, E., Khaksari, M., Mazzoleni, L. R., & Minakata, D. (2018). Transformations of dissolved organic matter induced by UV photolysis, Hydroxyl radicals, chlorine radicals, and sulfate radicals in aqueous-phase UV-Based advanced oxidation processes. *Water Research, 135*, 22–30. https://doi.org/10.1016/j.watres.2018.02.013.
28. Wang, J., Deng, J., Du, E., & Guo, H. (2022). Reevaluation of radical-induced differentiation in UV-based advanced oxidation processes (UV/hydrogen peroxide, UV/peroxydisulfate, and UV/chlorine) for metronidazole removal: Kinetics, mechanism, toxicity variation, and DFT

studies. *Separation and Purification Technology, 301*, 121905. https://doi.org/10.1016/j.seppur.2022.121905.

29. Xu, Y. (2001). Comparative studies of the Fe3+/2+–UV, H_2O_2–UV, TiO_2–UV/vis systems for the decolorization of a textile dye X-3B in water. *Chemosphere, 43*(8), 1103–1107. https://doi.org/10.1016/S0045-6535(00)00181-7.
30. Yang, W., Zhou, H., & Cicek, N. (2014). Treatment of organic micropollutants in water and wastewater by UV-based processes: A literature review. *Critical Reviews in Environmental Science and Technology, 44*(13), 1443–1476. https://doi.org/10.1080/10643389.2013.770440.
31. Yin, H., Zhang, Q., Su, Y., Tang, Y., & Zhou, M. (2021). A novel UV based advanced oxidation process with electrochemical co-generation of chlorine and H_2O_2 for carbamazepine abatement: Better performance, lower energy consumption and less DBPs formation. *Chemical Engineering Journal, 425*, 131857. https://doi.org/10.1016/j.cej.2021.131857.
32. Yu, L., Mehrabani-Zeinabad, M., Achari, G., & Langford, C. H. (2016). Application of UV based advanced oxidation to treat sulfolane in an aqueous medium. *Chemosphere, 160*, 155–161. https://doi.org/10.1016/j.chemosphere.2016.06.088.
33. Zhang, Y., Zhao, Y. G., Maqbool, F., & Hu, Y. (2022). Removal of antibiotics pollutants in wastewater by UV-based advanced oxidation processes: Influence of water matrix components, processes optimization and application: A review. *Journal of Water Process Engineering, 45*, 102496. https://doi.org/10.1016/j.jwpe.2022.102496.

INDEX

D

E

F

M

N

O

P

Q

R

S

T

Printed in the United States
by Baker & Taylor Publisher Services